IRISH PUB FOOD

CLASSIC PUB FARE THAT CAPTURES THE ESSENCE OF IRELAND

Publications International, Ltd.

Louis Weber, CEO
Publications International, Ltd.
8140 Lehigh Ave
Morton Grove, IL 60053

Pictured on the front cover: Guinness Lemb Stew (page 70).

Pictured on the back cover (clockwise from bottom): Shepherd's Pie Stuffed Potatoes (page 134), Curly Curry Chips (page 40), Open-Faced Lamb Sandwiches (page 93) and Mussels in Beer Broth (page 25).

ISBN: 978-1-63938-671-0

Manufactured in China.

8 7 6 5 4 3 2 1

Microwave Cooking: Microwave ovens vary in wattage. Use the cooking times as guidelines and check for doneness before adding more time.

WARNING: Food preparation, baking and cooking involve inherent dangers: misuse of electric products, sharp electric tools, boiling water, hot stoves, allergic reactions, foodborne illnesses and the like, pose numerous potential risks. Publications International, Ltd. (PIL) assumes no responsibility or liability for any damages you may experience as a result of following recipes, instructions, tips or advice in this publication.

While we hope this publication helps you find new ways to eat delicious foods, you may not always achieve the results desired due to variations in ingredients, cooking temperatures, typos, errors, omissions or individual cooking abilities.

Let's get social!

@Publications_International

@PublicationsInternational

www.pilbooks.com

CONTENTS

WEEKEND BRUNCH

Irish Porridge with Berry Compote

MAKES 4 SERVINGS

- 4 cups plus 1 tablespoon water, divided
- ½ teaspoon salt
- 1 cup steel-cut oats
- ½ teaspoon ground cinnamon
- ⅓ cup half-and-half
- ¼ cup packed brown sugar
- 1 cup fresh strawberries, hulled and quartered
- 1 container (6 ounces) fresh blackberries
- 1 container (6 ounces) fresh blueberries
- 3 tablespoons granulated sugar

1. Bring 4 cups water and salt to a boil in medium saucepan over medium-high heat. Whisk in oats and cinnamon. Reduce heat to medium; simmer, uncovered, about 40 minutes or until water is absorbed and oats are tender. Remove from heat; stir in half-and-half and brown sugar.
2. Meanwhile, combine strawberries, blackberries, blueberries, granulated sugar and remaining 1 tablespoon water in small saucepan; bring to a simmer over medium heat. Cook 8 to 9 minutes or until berries are tender but still hold their shape, stirring occasionally.
3. Divide porridge among four bowls; top with berry compote.

Waffled Breakfast Hash with Smoked Trout

MAKES 4 SERVINGS

- 1¼ pounds russet potatoes, peeled and cut into ½-inch pieces
- ½ small red onion, finely diced
- 1 small red bell pepper, cut into ½-inch pieces
- ¼ cup flaked smoked trout
- ⅓ cup sliced green onions
- 1 egg, lightly beaten
- 2 tablespoons vegetable oil
- 2 teaspoons cornstarch
- ½ teaspoon salt
- ¼ teaspoon black pepper
- 4 eggs, fried (optional)

1. Preheat classic waffle maker to medium-high heat. Set wire rack on baking sheet.
2. Place potatoes in large saucepan; add water to cover by 1 inch. Bring to a boil over high heat. Reduce heat to medium-low; partially cover and simmer 6 to 8 minutes or until potatoes are tender. Drain and run under cold water; transfer to large bowl.
3. Mash potatoes coarsely with fork or potato masher. Add onion, bell pepper, trout, green onions, beaten egg, oil, cornstarch, salt and black pepper; stir until blended.
4. Place 1 cup potato mixture in center of waffle maker. Close lid firmly; cook about 5 minutes or until waffle is golden brown and crisp. Remove to wire rack; tent with foil to keep warm. Repeat with remaining potato mixture.
5. Serve hash with fried eggs, if desired.

Cheddar and Leek Strata

MAKES 12 SERVINGS

- 8 eggs
- 2 cups milk
- ½ cup porter ale or stout
- 2 cloves garlic, minced
- ¼ teaspoon salt
- ¼ teaspoon black pepper
- 1 loaf (16 ounces) sourdough bread, cut into ½-inch cubes
- 2 small leeks, coarsely chopped
- 1 red bell pepper, chopped
- 1½ cups (6 ounces) shredded Swiss cheese
- 1½ cups (6 ounces) shredded sharp Cheddar cheese

1. Spray 13×9-inch baking dish with nonstick cooking spray. Whisk eggs, milk, ale, garlic, salt and black pepper in large bowl until well blended.
2. Spread half of bread cubes in prepared baking dish; sprinkle with half of leeks and half of bell pepper. Top with ¾ cup Swiss and ¾ cup Cheddar. Repeat layers. Pour egg mixture evenly over top.
3. Cover tightly with plastic wrap or foil. Weigh down top of strata with slightly smaller baking dish; refrigerate at least 2 hours or overnight.
4. Preheat oven to 350°F. Bake strata, uncovered, 40 to 45 minutes or until center is set. Serve immediately.

Irish Whiskey Cured Salmon

MAKES 6 TO 8 SERVINGS

- 1 skin-on salmon fillet (1¾ pounds), pin bones removed
- 2 tablespoons Irish whiskey
- ⅓ cup packed dark brown sugar
- 3 tablespoons salt
- Black bread or Irish soda bread (optional)
- Fresh dill, crème fraîche, thinly sliced red onion and/or capers (optional)

1. Line rimmed baking sheet with plastic wrap. Rinse salmon and pat dry with paper towels. Arrange salmon, skin side down, on prepared baking sheet; brush with whiskey.
2. Combine brown sugar and salt in small bowl; rub mixture over salmon. Wrap plastic wrap securely around salmon. Top with another sheet of plastic wrap.
3. Place second baking sheet on top of salmon, then place heavy skillet or several cans on top to weigh it down. Refrigerate salmon at least 48 hours or up to 72 hours.
4. Remove top baking sheet. Unwrap salmon and rinse under cold running water to remove any remaining salt mixture. Pat dry with paper towels. Cut salmon into very thin slices; serve with bread and assorted toppings, if desired. Refrigerate leftover salmon up to 2 days.

TIP

Ask your fishmonger to remove the pin bones when purchasing the salmon. (Often this is already done, or you can remove the pin bones at home with tweezers.)

Bacon and Potato Quiche

MAKES 8 SERVINGS

- **1 refrigerated pie crust (half of 14-ounce package)**
- **12 ounces thick-cut bacon, cut into ½-inch pieces**
- **8 ounces Yukon Gold potatoes, peeled and cut into ¼-inch pieces**
- **½ medium onion, chopped**
- **½ teaspoon chopped fresh thyme**
- **1½ cups half-and-half**
- **4 eggs**
- **½ teaspoon salt**
- **½ teaspoon black pepper**
- **¾ cup (3 ounces) shredded Dubliner cheese**
- **2 tablespoons chopped fresh chives**

1. Preheat oven to 450°F. Line baking sheet with foil.
2. Roll out pie crust into 12-inch circle on floured surface. Line 9-inch pie plate with crust, pressing firmly against bottom and up side of plate. Trim crust to leave 1-inch overhang; fold under and flute edge. Prick bottom of crust with fork. Bake about 8 minutes or until lightly browned. Remove to wire rack to cool slightly. *Reduce oven temperature to 375°F.*
3. Cook bacon in large skillet over medium heat about 10 minutes or until crisp, stirring occasionally. Drain on paper towel-lined plate. Drain all but 1 tablespoon drippings from skillet. Add potatoes, onion and thyme to skillet; cook about 10 minutes or until vegetables are tender, stirring occasionally.
4. Place pie plate on prepared baking sheet. Whisk half-and-half, eggs, salt and pepper in medium bowl until well blended. Sprinkle cheese evenly over bottom of crust; top with vegetable mixture and bacon. Pour in egg mixture; sprinkle with chives.
5. Bake 35 to 40 minutes or until quiche is set and knife inserted into center comes out clean. Let stand 10 minutes before slicing.

Oatmeal Pecan Pancakes

MAKES 4 SERVINGS

- 1¼ to 1½ cups milk, divided
- ½ cup old-fashioned oats
- ⅔ cup all-purpose flour
- ⅓ cup whole wheat flour
- 2½ tablespoons packed brown sugar
- 2 teaspoons baking powder
- ½ teaspoon baking soda
- ¼ teaspoon salt
- 1 egg
- 2 tablespoons butter, melted, plus additional for serving
- ½ cup chopped toasted pecans*
- Maple syrup

***To toast pecans, cook in small skillet over medium-low heat 4 to 5 minutes or until lightly browned and fragrant, stirring frequently.**

1. Bring ½ cup milk to a simmer in small saucepan. Stir in oats. Remove from heat; let stand 10 minutes.
2. Combine all-purpose flour, whole wheat flour, brown sugar, baking powder, baking soda and salt in large bowl; mix well.
3. Beat egg and melted butter in medium bowl until blended. Stir in oatmeal and ¾ cup milk. Add to flour mixture; stir just until blended. *Do not beat.* If batter is too thick, thin with remaining ¼ cup milk, 1 tablespoon at a time. Stir in pecans.
4. Lightly grease large skillet or griddle; heat over medium heat. Pour batter into skillet by ¼ cupfuls; flatten slightly. Cook 2 minutes or until tops are bubbly and bottoms are golden brown. Turn and cook 2 minutes or until golden brown. Serve with maple syrup and additional butter, if desired.

Mini Cream Scones

MAKES 16 SCONES

- 2 cups all-purpose flour
- ⅓ cup sugar
- 2 teaspoons baking powder
- ½ teaspoon salt
- ½ cup (1 stick) cold butter, cut into pieces
- ½ cup plus 2 tablespoons whipping cream
- 2 eggs, divided
- 1 teaspoon water

1. Preheat oven to 425°F. Line baking sheet with parchment paper.
2. Combine flour, sugar, baking powder and salt in large bow; mix well. Cut in butter with pastry blender until mixture resembles coarse crumbs.
3. Beat cream and 1 egg in small bowl until well blended. Add to flour mixture; stir just until blended.
4. Turn dough out onto lightly floured surface. Divide dough in half; shape into two ½-inch-thick rounds. Cut each round into eight wedges; place on prepared baking sheet. Beat remaining 1 egg and water in small bowl; brush over scones.
5. Bake about 10 minutes or until tops are firm and lightly browned. Remove to wire rack to cool slightly; serve warm or at room temperature.

Corned Beef Hash

MAKES 4 SERVINGS

- 2 large russet potatoes, peeled and cut into ½-inch pieces
- ½ teaspoon salt
- ¼ teaspoon black pepper
- ¼ cup (½ stick) butter
- 1 cup chopped onion
- 8 ounces corned beef, finely chopped
- 1 tablespoon horseradish
- 4 eggs

1. Place potatoes in large skillet; add water to cover. Bring to a boil over high heat. Reduce heat to low; cook 6 minutes. (Potatoes will be firm.) Drain potatoes; transfer to medium bowl. Season with salt and pepper.
2. Melt butter in same skillet over medium heat. Add onion; cook and stir 5 minutes. Add corned beef, horseradish and potatoes; mix well. Press mixture with spatula to flatten.
3. Reduce heat to low; cook 10 to 15 minutes. Turn hash in large pieces; press down and cook 10 to 15 minutes or until bottom is well browned.
4. Meanwhile, bring 1 inch of water to a simmer in small saucepan. Break 1 egg into shallow dish; carefully slide into water. Cook 5 minutes or until white is opaque. Remove with slotted spoon to plate; keep warm. Repeat with remaining eggs.
5. Top each serving of hash with egg. Serve immediately.

Oatmeal Brûlée with Raspberry Sauce

MAKES 4 SERVINGS

- 4¼ cups water, divided
- ½ teaspoon salt
- 3 cups old-fashioned oats
- 1 cup whipping cream
- ½ teaspoon vanilla
- ¾ cup granulated sugar, divided
- 3 egg yolks
- 6 ounces frozen sweetened raspberries
- ½ teaspoon orange extract
- 2 tablespoons packed brown sugar

1. Preheat oven to 300°F. Line baking sheet with foil. Bring 4 cups water and salt to a boil in medium saucepan over high heat. Add oats; cook over low heat 3 to 5 minutes or until water is absorbed and oats are tender, stirring occasionally. Divide oatmeal among four large ramekins or ovenproof bowls. Place on prepared baking sheet.
2. Bring cream to a simmer in small saucepan over medium heat. *Do not boil.* Remove from heat; stir in vanilla. Whisk ¼ cup granulated sugar and egg yolks in medium bowl. Slowly pour about ½ cup hot cream into egg mixture, whisking constantly. Stir egg mixture back into saucepan with cream, whisking until well blended. Pour cream mixture evenly over oatmeal in ramekins.
3. Bake 35 minutes or until almost set. Remove from oven. *Turn oven to broil.*
4. Meanwhile, prepare sauce. Combine raspberries, remaining ½ cup granulated sugar, ¼ cup water and orange extract in blender or food processor; blend until smooth. Strain sauce into pitcher or bowl.
5. Sprinkle ½ tablespoon brown sugar evenly over each custard. Broil 3 to 5 minutes or until sugar melts and turns golden brown. Cool 5 to 10 minutes before serving. Serve with raspberry sauce.

SIMPLE CHEDDAR BISCUITS

MAKES 9 BISCUITS

- 1 cup all-purpose flour
- 2 teaspoons baking powder
- 1 teaspoon sugar
- ½ teaspoon salt
- ¼ cup (½) stick butter, cut into small pieces
- 1 cup (4 ounces) grated sharp Cheddar cheese
- ½ cup milk

1. Preheat oven to 425°F. Line baking sheet with parchment paper.
2. Combine flour, baking powder, sugar and salt in medium bowl; mix well. Cut in butter with pastry blender until mixture resembles coarse crumbs. Add cheese; stir until blended. Add milk; stir until soft dough forms.
3. Turn out dough onto lightly floured surface; knead several times until smooth, sprinkling with additional flour as necessary. Pat dough into 6-inch square; cut into nine (2-inch) squares. Place 2 inches apart on prepared baking sheet.
4. Bake 10 to 12 minutes or until golden brown. Cool on baking sheet 2 minutes; serve warm.

DATE-NUT GRANOLA

MAKES 6 CUPS

- 2 cups old-fashioned oats
- 2 cups barley flakes
- 1 cup sliced almonds
- ⅓ cup vegetable oil
- ⅓ cup honey
- 1 teaspoon vanilla
- 1 cup chopped dates

1. Preheat oven to 350°F. Spray 13×9-inch baking pan with nonstick cooking spray.
2. Combine oats, barley flakes and almonds in large bowl. Whisk oil, honey and vanilla in small bowl until well blended. Pour honey mixture over oat mixture; stir until blended. Spread in prepared pan.
3. Bake 25 minutes or until toasted, stirring frequently after 10 minutes. Stir in dates while granola is still hot. Cool completely. Store tightly covered.

APPETIZERS & SNACKS

MUSSELS IN BEER BROTH

MAKES 4 SERVINGS

- 2 tablespoons olive oil
- ⅓ cup chopped shallots
- 4 cloves garlic, minced
- 2 cups pale ale
- 1 can (about 14 ounces) diced tomatoes
- ¼ cup chopped fresh parsley
- 1 tablespoon chopped fresh thyme
- ½ teaspoon salt
- ¼ teaspoon red pepper flakes
- 3 pounds mussels, scrubbed and debearded
- French bread (optional)

1. Heat oil in large saucepan or Dutch oven over medium-high heat. Add shallots and garlic; cook and stir 3 minutes or until softened. Stir in ale, tomatoes, parsley, thyme, salt and red pepper flakes; bring to a boil.
2. Add mussels. Reduce heat to low; cover and simmer 5 to 7 minutes or until mussels open. Discard any unopened mussels. Serve with bread, if desired.

HAM AND RICE CROQUETTES

MAKES 6 SERVINGS

- ⅔ cup water
- ⅓ cup uncooked rice
- 2 ounces ham, finely chopped
- 1 egg yolk, beaten
- 1 tablespoons capers, drained and rinsed
- ⅛ teaspoon salt
- ⅛ teaspoon dried oregano
- ⅛ teaspoon black pepper
- ⅔ cup fresh bread crumbs
- 1 to 2 tablespoons butter
- 1 tablespoon olive oil

1. Bring water to a boil in small saucepan over high heat. Stir in rice. Reduce heat to low; cover and simmer about 14 minutes or until rice is tender and water has been absorbed. Transfer rice to medium bowl; cool until almost room temperature.
2. Add ham, egg yolk, capers, salt, oregano and pepper to rice; mix well. Spread bread crumbs on large plate. Shape rice mixture into 18 (1¼-inch) balls. Flatten balls slightly; roll in bread crumbs to coat. Place on baking sheet; refigerate 15 to 30 minutes or until firm.
3. Heat 1 tablespoon butter and oil in large skillet over medium-high heat until butter melts. Add half of croquettes; cook 2 to 3 minutes or until golden brown. Turn and cook 1 to 2 minutes or until golden brown. Remove to plate; keep warm. Repeat with remaining croquettes, adding additional butter to skillet if necessary. Serve immediately.

BEER-BATTERED SHRIMP

MAKES 4 TO 6 SERVINGS

- ¾ cup mayonnaise
- ⅓ cup cocktail sauce
- 1¼ cups all-purpose flour
- 1 teaspoon baking powder
- ½ teaspoon sweet paprika
- ½ teaspoon salt
- 1 bottle (12 ounces) lager, as needed
- Vegetable oil for frying
- 1½ pounds large raw shrimp, peeled and deveined

1. Combine mayonnaise and cocktail sauce in medium bowl; mix well. Cover and refrigerate until ready to use.
2. Whisk flour, baking powder, paprika and salt in medium bowl until blended. Whisk in enough lager to make thick batter. Cover and let stand at room temperature 2 to 4 hours.
3. Preheat oven to 200°F. Line large baking sheet with paper towels. Fill large saucepan half full with oil. Heat over medium-high heat to 350°F; adjust heat to maintain temperature.
4. Working in batches, dip shrimp into batter, letting excess drip back into bowl. Carefully add shrimp to oil; cook 2½ minutes or until golden brown. Remove to prepared baking sheet with slotted spoon; keep warm in oven while cooking remaining shrimp. Serve warm with dip.

Wild Mushroom Flatbread

MAKES ABOUT 8 SERVINGS

- 1 package (about 14 ounces) refrigerated pizza dough
- 1 tablespoon olive oil
- 1 package (4 ounces) sliced cremini mushrooms
- 1 package (4 ounces) sliced shiitake mushrooms
- 1 shallot, thinly sliced
- 2 cloves garlic, minced
- ½ teaspoon salt
- ¾ cup (3 ounces) grated Gruyère cheese
- 2 teaspoons chopped fresh thyme

1. Preheat oven to 400°F. Line baking sheet with parchment paper; spray with nonstick cooking spray.
2. Roll out dough into 15×10-inch rectangle on lightly floured surface. Transfer to prepared baking sheet. Bake 10 minutes.
3. Meanwhile, heat oil in large skillet over medium-high heat. Add mushrooms; cook and stir 5 minutes. Add shallot and garlic; cook and stir 5 minutes or until tender. Season with salt.
4. Spread mushroom mixture evenly over crust; sprinkle with cheese and thyme.
5. Bake 8 minutes or until cheese is melted.

CALAMARI WITH TARTAR SAUCE

MAKES 2 TO 3 SERVINGS

- Tartar Sauce (recipe follows)
- 1 pound cleaned squid (body tubes, tentacles or a combination), rinsed and patted dry
- ¾ cup plain dry bread crumbs
- 1 egg
- 1 tablespoon milk
- Vegetable oil
- Lemon wedges (optional)

1. Prepare Tartar Sauce. Line baking sheet with waxed paper or parchment paper. Cut squid into ¼-inch rings.
2. Place bread crumbs in medium bowl. Beat egg and milk in another medium bowl. Add squid; stir to coat. Transfer squid to bowl with bread crumbs; toss to coat. Place on prepared baking sheet; refrigerate 15 minutes.
3. Heat 1½ inches oil in large saucepan over medium-high heat to 350°F; adjust heat to maintain temperature. Cook squid in batches, 8 to 10 pieces at a time, 45 seconds or until golden brown. (Squid will pop and spatter during frying; do not stand too close to saucepan.) *Do not overcook.* Remove to paper towel-lined plate.
4. Serve immediately with Tartar Sauce and lemon wedges, if desired.

TARTAR SAUCE

MAKES ABOUT 1⅓ CUPS

- 1⅓ cups mayonnaise
- 2 tablespoons chopped fresh Italian parsley
- 1 green onion, thinly sliced
- 1 tablespoon drained capers, minced
- 1 small sweet gherkin or pickle, minced

Combine mayonnaise, parsley, green onion, capers and gherkin in small bowl; mix well. Cover and refrigerate until ready to serve.

Sausage Rolls

MAKES 4 SERVINGS

- 8 ounces ground pork
- ¼ cup finely chopped onion
- ½ teaspoon coarse salt
- 1 teaspoon minced garlic
- ½ teaspoon dried thyme
- ½ teaspoon dried basil
- ¼ teaspoon dried marjoram
- ¼ teaspoon black pepper
- 1 sheet frozen puff pastry (half of 17-ounce package), thawed
- 1 egg, beaten

1. Preheat oven to 400°F. Line baking sheet with parchment paper.
2. Combine pork, onion, salt, garlic, thyme, basil, marjoram and pepper in medium bowl; mix well.
3. Place puff pastry on floured surface; cut lengthwise into three strips at seams. Roll each third into 10×4½-inch rectangle. Shape one third of pork mixture into 10-inch log; arrange log along top edge of one pastry rectangle. Brush bottom ½ inch of rectangle with egg. Roll pastry down around pork; press to seal.
4. Cut roll crosswise into four pieces; place seam side down on prepared baking sheet. Repeat with remaining puff pastry and pork mixture. Brush top of each roll with egg.
5. Bake about 25 minutes or until sausage is cooked through and pastry is golden brown and puffed. Remove to wire rack to cool 10 minutes. Serve warm.

FRIED ZUCCHINI

MAKES 4 SERVINGS

- Lemon Aioli (recipe follows)
- Vegetable oil for frying
- ¾ to 1 cup soda water
- ½ cup all-purpose flour
- ¼ cup cornstarch
- ½ teaspoon coarse salt
- ¼ teaspoon garlic powder
- ¼ teaspoon dried oregano
- ¼ teaspoon black pepper
- 3 cups panko bread crumbs
- 1½ pounds medium zucchini (6 to 8 inches long), ends trimmed, cut lengthwise into ¼-inch-thick slices
- ¼ cup grated Parmesan or Romano cheese
- Chopped fresh parsley
- Lemon wedges

1. Prepare Lemon Aioli; cover and refrigerate until ready to use.
2. Line baking sheet with paper towels; set aside. Heat 2 inches oil in large saucepan or Dutch oven over medium-high heat to 350°F; adjust heat to maintain temperature.
3. Meanwhile, pour ¾ cup soda water into large bowl. Combine flour, cornstarch, salt, garlic powder, oregano and pepper in small bowl; mix well. Slowly whisk flour mixture into soda water just until blended. Add additional soda water if necessary to reach consistency of thin pancake batter. Place panko in medium bowl.
4. Working with one at a time, dip zucchini slices into batter to coat, letting excess batter drip back into bowl. Add to bowl with panko; press panko into zucchini slices to coat both sides completely. Place zucchini on prepared baking sheet.
5. Cook zucchini in batches 3 to 4 minutes or until golden brown. (Return oil to 350°F between batches.) Remove to paper towel-lined plate. Sprinkle with cheese and parsley; serve with Lemon Aioli and lemon wedges.

LEMON AIOLI

Combine ½ cup mayonnaise, 2 tablespoons lemon juice, 1 tablespoon chopped fresh Italian parsley and 1 teaspoon minced garlic in small bowl; mix well. Season with salt and pepper.

Bacon and Cheese Rarebit

MAKES 6 SERVINGS

- 1½ tablespoons butter
- ½ cup lager (not dark beer)
- 2 teaspoons Dijon mustard
- 2 teaspoons Worcestershire sauce
- ⅛ teaspoon ground red pepper
- 2 cups (8 ounces) shredded American cheese
- 1½ cups (6 ounces) shredded sharp Cheddar cheese
- 1 small loaf (8 ounces) egg bread or brioche, cut into 6 (1-inch-thick) slices
- 12 large slices tomato
- 12 slices bacon, crisp-cooked
- Finely chopped fresh thyme (optional)

1. Preheat broiler. Line baking sheet with foil.
2. Melt butter in medium saucepan over medium-low heat. Stir in lager, mustard, Worcestershire sauce and red pepper; cook until heated through, stirring occasionally. Gradually add cheeses, stirring constantly until melted. Remove from heat; cover and keep warm.
3. Broil bread slices until golden brown. Arrange on prepared baking sheet. Top each slice with tomato and bacon. Spoon about ¼ cup cheese sauce evenly over each serving.
4. Broil 4 to 5 inches from heat just until cheese sauce begins to brown. Garnish with thyme.

Curly Curry Chips

MAKES 4 SERVINGS

- 4 small or 2 large russet potatoes, peeled
- 2 teaspoons vegetable oil
- ¾ teaspoon salt, divided
- 1 tablespoon butter
- ¼ cup finely chopped onion
- 1 tablespoon all-purpose flour
- 1 tablespoon curry powder
- 1 cup reduced-sodium vegetable broth

1. Preheat oven to 450°F. Line large baking sheet with parchment paper. Spiral potatoes with thick spiral blade of spiralizer. (See Note.)
2. Spread potatoes on prepared baking sheet; drizzle with oil. Bake 30 to 35 minutes or until golden brown and crisp, turning once. Sprinkle with ½ teaspoon salt.
3. Meanwhile, melt butter in small saucepan over medium-high heat. Add onion; cook and stir 3 minutes or until softened. Whisk in flour and curry powder until well blended; cook 1 minute, stirring constantly. Add broth in thin steady stream, whisking constantly.
4. Reduce heat to medium; cook about 10 minutes or until thick. Taste and add ¼ teaspoon salt, if desired. For smoother sauce, cool slightly and purée in blender or food processor. Serve with potatoes.

NOTE

If you don't have a spiralizer, make regular oven fries instead. Peel potatoes and cut lengthwise into ¼-inch strips. Place in colander; rinse under cold water 2 minutes. Pat dry with paper towels. Toss with 2 teaspoons oil in medium bowl until coated. Spread in single layer on baking sheet. Bake about 25 minutes or until golden brown and crisp, turning once.

SCOTCH EGGS

MAKES 8 SERVINGS

- 10 eggs, divided
- 2 tablespoons vegetable oil
- 1½ cups panko bread crumbs
- 1 pound bulk breakfast sausage
- ¼ cup thinly sliced green onions
- ¾ cup all-purpose flour
- 2 tablespoons whole grain mustard

1. Preheat oven to 400°F. Line large baking sheet with foil.
2. Place 8 eggs in large saucepan filled with cold water; cover and bring to a boil over medium-high heat. Turn off heat; let stand 10 minutes. Run eggs under cool water to stop cooking. When cool enough to handle, carefully crack and peel eggs.
3. Meanwhile, heat oil in medium skillet over medium heat. Add panko; cook about 8 minutes or until toasted and golden brown, stirring occasionally. Remove to medium bowl; let cool.
4. Combine sausage and green onions in medium bowl. Place flour in shallow bowl. Lightly beat remaining 2 eggs and mustard in another shallow bowl.
5. Scoop out one eighth of sausage mixture; press flat in palm of your hand. Place 1 cooked egg in center of mixture and wrap sausage around it. Gently roll between your hands until sausage completely encloses egg. Coat sausage-wrapped egg with flour, shaking off excess. Dip in egg-mustard mixture; roll in panko to coat. Place on prepared baking sheet. Repeat with remaining eggs and sausage.
6. Bake 16 to 18 minutes or until sausage is cooked through. Drain well on paper towel-lined plate. Serve immediately.

Rhubarb Chutney

MAKES ABOUT 2 CUPS

- 1 cup coarsely chopped peeled apple
- ½ cup sugar
- ¼ cup water
- ¼ cup dark raisins
- 1 teaspoon grated lemon peel
- 2 cups sliced fresh rhubarb (½-inch pieces)
- 3 tablespoons coarsely chopped pecans
- 2 to 3 teaspoons white vinegar
- ¾ teaspoon ground cinnamon (optional)

1. Combine apple, sugar, water, raisins and lemon peel in medium saucepan; cook over medium heat until sugar is dissolved, stirring constantly. Reduce heat to low; cook, uncovered, about 5 minutes or until apple is almost tender.
2. Stir in rhubarb and pecans; bring to a boil over high heat. Reduce heat to low; cook 8 to 10 minutes or until slightly thickened, stirring occasionally. Stir in vinegar and cinnamon, if desired, during last 2 to 3 minutes of cooking.
3. Remove from heat; cool to room temperature. Cover and refrigerate until ready to serve. Serve with cheese and crackers, roast chicken or pork.

CRISPY OVEN-FRIED MUSHROOMS

MAKES 4 SERVINGS

- ¼ cup panko breadcrumbs
- 2 teaspoons garlic powder
- 1½ teaspoons poultry seasoning
- 1½ teaspoons celery seed
- 1½ teaspoons paprika
- 1 teaspoon onion powder
- ½ teaspoon sage
- ½ teaspoon thyme
- ½ cup all-purpose flour
- 1 cup buttermilk
- 2 cups oyster mushrooms
- Blue cheese dressing

1. Preheat oven to 425°F. Line baking sheet with parchment paper.
2. Combine panko, garlic powder, poultry seasoning, celery seed, paprika, onion powder, sage and thyme in medium bowl; mix well. Place flour in large bowl; slowly whisk in buttermilk until well blended and no longer lumpy.
3. Use fork to dip mushrooms into buttermilk mixture, coating completely. Roll in panko mixture to coat.
4. Place mushrooms on prepared baking sheet; spray with nonstick cooking spray.
5. Bake 15 minutes; turn and bake 5 to 10 minutes or until mushrooms are well browned. Serve with dressing.

TIP

To cook the mushrooms in an air fryer, preheat the air fryer to 390°F and line the basket with parchment paper. Prepare the mushrooms as directed, then cook in batches 6 to 8 minutes or until browned, shaking occasionally during cooking.

BIG SALADS

Beet and Arugula Salad

MAKES 6 SERVINGS

- 8 medium beets (5 to 6 ounces each)
- ⅓ cup red wine vinegar
- ¾ teaspoon salt
- ½ teaspoon black pepper
- 3 tablespoons extra virgin olive oil
- 1 package (5 ounces) baby arugula
- 1 package (4 ounces) crumbled goat cheese with garlic and herbs

1. Place beets in large saucepan; add water to cover by 2 inches. Bring to a boil over medium-high heat. Reduce heat to medium-low; cover and simmer 30 minutes or until beets can be easily pierced with tip of knife. Drain well; set aside until cool enough to handle.
2. Meanwhile, whisk vinegar, salt and pepper in large bowl. Slowly whisk in oil in thin, steady stream until well blended. Remove 3 tablespoons dressing to medium bowl.
3. Peel beets and cut into wedges. Add warm beets to large bowl; toss to coat with dressing. Add arugula to medium bowl; toss gently to coat with dressing. Place arugula on platter or plates, top with beets and cheese.

Roasted Brussels Sprouts Salad

MAKES 6 SERVINGS

BRUSSELS SPROUTS

- 1 pound Brussels sprouts, trimmed and halved
- 2 tablespoons olive oil
- ½ teaspoon salt

SALAD

- 2 cups coarsely chopped baby kale
- 2 cups coarsely chopped romaine lettuce
- 1½ cups glazed pecans*
- 1 cup halved red grapes
- 1 cup diced cucumbers
- ½ cup dried cranberries
- ½ cup fresh blueberries
- ½ cup chopped red onion
- ¼ cup toasted pumpkin seeds (pepitas)
- 1 container (4 ounces) crumbled goat cheese

DRESSING

- ½ cup olive oil
- 6 tablespoons balsamic vinegar
- 6 tablespoons strawberry jam
- 2 teaspoons Dijon mustard
- 1 teaspoon salt

**Glazed pecans may be found in the produce section of the supermarket with other salad toppings, or they may be found in the snack aisle.*

1. For Brussels sprouts, preheat oven to 400°F. Spray baking sheet with nonstick cooking spray.
2. Combine Brussels sprouts, 2 tablespoons oil and ½ teaspoon salt in medium bowl; toss to coat. Arrange Brussels sprouts in single layer, cut sides down, on prepared baking sheet. Roast 20 minutes or until tender and browned, stirring once halfway through roasting. Cool completely on baking sheet.
3. For salad, combine kale, lettuce, pecans, grapes, cucumbers, cranberries, blueberries, onion and pumpkin seeds in large bowl. Top with Brussels sprouts and cheese.
4. For dressing, whisk ½ cup oil, vinegar, jam, mustard and 1 teaspoon salt in small bowl until well blended. Pour dressing over salad; toss gently to coat.

WARM POTATO SALAD

MAKES 6 TO 8 SERVINGS

- 2 pounds fingerling potatoes, unpeeled
- 3 slices thick-cut bacon, cut into ½-inch pieces
- 1 small onion, diced
- 2 tablespoons olive oil
- ¼ cup cider vinegar
- 2 tablespoons capers, drained
- 1 tablespoon Dijon mustard
- ¾ teaspoon salt
- ¼ teaspoon black pepper
- ⅓ cup chopped fresh parsley

1. Place potatoes in large saucepan; add cold water to cover by 2 inches. Bring to a boil over high heat. Reduce heat to medium; cook 10 to 12 minutes or just until potatoes are tender when pierced with tip of small knife.
2. Drain potatoes; let stand until cool enough to handle. Meanwhile, dry out saucepan with paper towels. Add bacon to saucepan; cook until crisp, stirring occasionally. Remove to paper towel-lined plate. Drain off all but 1 tablespoon drippings.
3. Add onion and oil to saucepan; cook 10 minutes or until onion begins to turn golden, stirring occasionally. Cut potatoes crosswise into ½-inch slices.
4. Add vinegar, capers, mustard, salt and pepper to saucepan; mix well. Remove from heat; stir in potatoes. Add parsley and bacon; stir gently to coat.

CHICKEN AND APPLE SALAD

MAKES 4 SERVINGS (1 CUP DRESSING)

DRESSING

- 5 tablespoons apple juice concentrate
- ¼ cup white balsamic vinegar
- 1 tablespoon lemon juice
- 1 tablespoon sugar
- 1 clove garlic, minced
- ½ teaspoon salt
- ½ teaspoon onion powder
- ¼ teaspoon ground ginger
- ¼ cup extra virgin olive oil

SALAD

- 12 cups mixed greens such as chopped romaine lettuce and spring greens
- 12 ounces thinly sliced cooked chicken
- 2 tomatoes, cut into wedges
- 1 package (about 3 ounces) dried apple chips
- ½ red onion, thinly sliced
- ½ cup crumbled gorgonzola or blue cheese
- ½ cup pecans, toasted*

**To toast pecans, cook in small skillet over medium-low heat 4 to 5 minutes or until fragrant, stirring frequently.*

1. For dressing, whisk apple juice concentrate, vinegar, lemon juice, sugar, garlic, salt, onion powder and ginger in small bowl until blended. Slowly whisk in oil in thin, steady stream until well blended.
2. For salad, divide greens among four serving bowls. Top with chicken, tomatoes, apple chips, onion, cheese and pecans.
3. Drizzle about 2 tablespoons dressing over each salad.

Quinoa Salad with Mustard Vinaigrette

MAKES 6 SERVINGS

- 1 cup tri-colored uncooked quinoa
- 2 cups vegetable broth
- 1 tablespoon chopped fresh rosemary
- 1 package (12 ounces) fresh haricots verts, cut in half
- 3 tablespoons olive oil
- 1 tablespoon Dijon mustard
- 1 teaspoon honey
- 1 tablespoon lemon juice
- ½ teaspoon salt
- ⅛ teaspoon black pepper
- ½ cup toasted pecan pieces*
- 1 container (4 ounces) crumbled goat cheese

***To toast pecans, cook in small skillet over medium-low heat 2 to 3 minutes or until fragrant, stirring frequently.**

1. Place quinoa in fine-mesh strainer; rinse well under cold running water.
2. Combine quinoa and broth in medium saucepan; bring to a boil over high heat. Reduce heat to low; cover and simmer 15 to 20 minutes or until quinoa is tender and broth is absorbed. Add rosemary and haricots verts during last 5 minutes of cooking. Remove from heat; cool to room temperature.
3. Meanwhile, whisk oil, mustard, honey, lemon juice, salt and pepper in small bowl until well blended.
4. Place quinoa mixture in large bowl. Add dressing and pecans; toss to coat. Sprinkle with goat cheese.

Crunchy Kale Salad

MAKES 6 SERVINGS

- ¼ cup cider vinegar
- ¼ cup extra virgin olive oil
- ¼ cup maple syrup
- 1 tablespoon lemon juice
- ½ tablespoon Dijon mustard
- ½ teaspoon salt
- ¼ teaspoon black pepper
- 10 cups chopped stemmed kale (about 1 large bunch)
- 2 cups shredded green cabbage
- ½ cup sliced almonds, toasted*

**To toast almonds, cook in small skillet over medium-low heat 2 to 3 minutes or until lightly browned and fragrant, stirring frequently.*

1. Whisk vinegar, oil, maple syrup, lemon juice, mustard, salt and pepper in small bowl or measuring cup until well blended.
2. Combine kale and cabbage in large bowl. Pour dressing over vegetables; massage kale with hands 3 to 4 minutes to soften. Stir in almonds just before serving.

Market Salad

MAKES 4 SERVINGS

- 3 eggs
- 4 cups mixed baby salad greens
- 2 cups green beans, cut into 1½-inch pieces, cooked and drained
- 4 slices thick-cut bacon, crisp-cooked and crumbled
- 1 tablespoon minced fresh basil, chives or Italian parsley
- 3 tablespoons olive oil
- 1 tablespoon red wine vinegar
- 1 teaspoon Dijon mustard
- ¼ teaspoon salt
- ¼ teaspoon black pepper

1. Place eggs in small saucepan; add water to cover. Bring to a boil over medium-high heat. Immediately remove from heat; cover and let stand 10 minutes. Drain and cool eggs to room temperature.
2. Combine salad greens, green beans, bacon and basil in large serving bowl. Peel and coarsely chop eggs; add to serving bowl. Whisk oil, vinegar, mustard, salt and pepper in small bowl until well blended. Drizzle dressing over salad; toss gently to coat.

Flank Steak and Roasted Vegetable Salad

MAKES 4 SERVINGS

- 1½ pounds asparagus spears, trimmed and cut into 2-inch pieces
- 1¾ cups baby carrots (8 ounces)
- 2 tablespoons olive oil, divided
- ¾ teaspoon salt, divided
- 1 teaspoon black pepper, divided
- 1 pound flank steak (1 inch thick)
- 2 tablespoons plus 1 teaspoon Dijon mustard, divided
- 1 tablespoon lemon juice
- 1 tablespoon water
- 1 teaspoon honey
- 6 cups mixed salad greens

1. Preheat oven to 400°F. Combine asparagus and carrots on baking sheet. Drizzle with 1 tablespoon oil; sprinkle with ¼ teaspoon salt and ¼ teaspoon pepper and toss to coat. Spread vegetables in single layer.
2. Roast 20 minutes or until vegetables are tender and lightly browned, stirring once. Meanwhile, rub both sides of steak with 2 tablespoons mustard; sprinkle with ¼ teaspoon salt and ½ teaspoon pepper. Place steak on rack in baking pan.
3. Roast steak 10 minutes for medium rare or until desired doneness, turning once. Remove to cutting board; let stand 5 minutes before cutting into thin slices across the grain.
4. Whisk lemon juice, remaining 1 tablespoon oil, 1 teaspoon mustard, ¼ teaspoon salt, ¼ teaspoon pepper, water and honey in large bowl until well blended. Drizzle 1 tablespoon dressing over vegetables in pan; stir to coat.
5. Add mixed greens to dressing in bowl; toss gently to coat. Divide greens among serving plates; top with steak and vegetables.

Green Bean Salad

MAKES 6 SERVINGS

- 1 pound fresh green beans, trimmed
- 2 tablespoons red wine vinegar
- 1 teaspoon honey
- 1 teaspoon Dijon mustard
- ½ teaspoon minced garlic
- ½ teaspoon salt
- ¼ teaspoon black pepper
- ¼ cup olive oil
- 1 red bell pepper, cut into thin strips
- 1 small red onion, cut into thin strips
- 2 tablespoons chopped fresh parsley
- Grated lemon peel (optional)

1. Fill large saucepan half full with salted water; bring to a boil over medium-high heat. Add beans; cook 4 to 5 minutes or until crisp-tender. Drain beans; transfer to bowl of ice water to stop cooking. When beans are cool, drain and pat dry.
2. Whisk vinegar, honey, mustard, garlic, salt and pepper in large bowl until well blended. Slowly whisk in oil in thin, steady stream until well blended.
3. Add beans, bell pepper, onion and parsley to dressing; toss to coat. Let stand at least 30 minutes before serving. Garnish with lemon peel.

TUNA SALAD NIÇOISE

MAKES 6 SERVINGS

HERB VINAIGRETTE

- ¼ cup white wine vinegar
- ¼ cup red wine vinegar
- ¼ cup chopped fresh basil
- 2 tablespoons chopped fresh chives
- 1 tablespoon Dijon mustard
- 2 cloves garlic, minced
- ½ teaspoon sugar
- ½ teaspoon salt
- ½ teaspoon black pepper
- 1 cup olive oil

SALAD

- 1 pound tuna steaks*
- 1½ pounds red potatoes, cut into 1-inch pieces
- 2 cups trimmed and halved green beans
- 8 cups mesclun greens or chopped romaine lettuce
- 3 hard-cooked eggs, cut into wedges
- ½ cup pitted Kalamata or black Niçoise olives
- 4 medium tomatoes, cut into wedges

**Or substitute 2 cans (6 ounces each) tuna, drained and flaked and skip steps 2 and 3.*

1. For vinaigrette, combine vinegars, basil, chives, mustard, garlic, sugar, salt and pepper in food processor or blender; process 30 seconds or until smooth. With motor running, slowly add oil; process until smooth. Pour dressing into jar; refrigerate until ready to use.
2. For salad, place tuna in baking dish. Pour ¼ cup dressing over tuna; turn to coat. Marinate in refrigerator 30 minutes.
3. Prepare grill for direct cooking over medium heat or preheat broiler. Remove tuna from marinade; discard marinade. Grill or broil tuna 4 minutes per side or until desired doneness. Remove to cutting board; let stand 5 minutes. Cut into thin slices.
4. Bring large saucepan of salted water to a boil over medium-high heat. Add potatoes; cook 5 minutes. Add beans; cook 5 minutes or until potatoes are tender and beans are crisp-tender. Drain and return vegetables to saucepan. Add ¼ cup dressing; stir to coat.
5. Divide greens among serving plates; top with potato and bean mixture. Arrange eggs, olives, tomatoes and tuna on top. Serve with remaining dressing.

Classic Irish Salad

MAKES 4 SERVINGS

DRESSING

- 3 tablespoons mayonnaise
- 1 tablespoon Dijon mustard
- 1 tablespoon canola oil
- 1 tablespoon cider vinegar
- 2 teaspoons sugar
- ¼ teaspoon salt
- ⅛ teaspoon black pepper

SALAD

- 6 cups torn romaine lettuce
- 2 cups baby arugula
- 1 large cucumber, halved lengthwise and thinly sliced
- 4 radishes, thinly sliced
- 3 tablespoons chopped fresh chives
- 2 hard-cooked eggs, cut into wedges
- 2 bottled pickled beets, quartered

1. For dressing, whisk mayonnaise, mustard, oil, vinegar, sugar, salt and pepper in small bowl until well blended.
2. For salad, combine romaine, arugula, cucumber, radishes and chives in large bowl. Divide among serving plates; top with eggs and beets. Serve dressing separately or drizzle over salads just before serving.

SOUPS & STEWS

Corned Beef and Cabbage Soup

MAKES ABOUT 8 SERVINGS

- 1 tablespoon vegetable oil
- 1 onion, chopped
- 2 stalks celery, chopped
- 2 carrots, chopped
- 2 cloves garlic, minced
- 4 to 5 cups coarsely chopped green cabbage (about half of 1 small head)
- 12 ounces unpeeled Yukon gold potatoes, chopped
- 4 cups beef broth
- 4 cups water
- ½ cup quick-cooking barley
- 1 teaspoon salt
- 1 teaspoon dried thyme
- ½ teaspoon black pepper
- ¼ teaspoon ground mustard
- 12 ounces corned beef (leftovers or deli corned beef, about 2½ cups), cut into ½-inch pieces

1. Heat oil in large saucepan or Dutch oven over medium-high heat. Add onion, celery and carrots; cook 5 minutes or until vegetables are softened, stirring occasionally. Add garlic; cook and stir 1 minute.
2. Stir in cabbage, potatoes, broth, water, barley, salt, thyme, pepper and mustard; bring to a boil. Reduce heat to medium-low; cook 20 minutes, stirring occasionally.
3. Stir in corned beef; cook 10 to 15 minutes or until potatoes are tender. Season with additional salt and pepper, if desired.

Guinness Lamb Stew

MAKES 6 SERVINGS

- 2 pounds boneless lamb stew meat (1- to 1½-inch pieces)
- 2 teaspoons salt, divided
- 1 teaspoon black pepper
- 2 tablespoons vegetable oil, divided
- 2 medium onions, chopped
- 2 cloves garlic, minced
- 1 tablespoon tomato paste
- 3 tablespoons all-purpose flour
- 2 cups beef broth
- 1 cup Guinness stout
- 2 large carrots, cut into ½-inch pieces
- 4 sprigs fresh thyme, plus additional for garnish
- 2 bay leaves
- 1½ pounds baking or yellow potatoes (about 3), peeled, cut lengthwise into quarters and crosswise into ½-inch pieces
- Chopped fresh parsley (optional)

1. Preheat oven to 325°F. Combine lamb, 1 teaspoon salt and pepper in medium bowl; toss to coat.
2. Heat 1 tablespoon oil in Dutch oven over medium-high heat. Add half of lamb in single layer (do not crowd); cook about 8 minutes or until well browned. Remove to plate. Repeat with remaining 1 tablespoon oil and lamb.
3. Add onions to Dutch oven; cook and stir about 4 minutes or until softened, scraping up browned bits from bottom of Dutch oven. Add garlic; cook and stir 1 minute. Add tomato paste; cook and stir 1 minute. Return lamb to Dutch oven; sprinkle with flour. Cook and stir 2 minutes or until flour is well incorporated. Add broth, Guinness, carrots, thyme, bay leaves and remaining 1 teaspoon salt; bring to a simmer, stirring occasionally.
4. Cover Dutch oven, leaving lid slightly open. Bake 1½ hours, stirring every 30 minutes. Stir in potatoes; cover with lid slightly ajar and bake 30 to 40 minutes or until lamb and potatoes are tender.
5. Remove thyme sprigs and bay leaves. Season with additional salt and pepper; garnish with parsley and additional thyme sprigs.

Creamy Onion Soup

MAKES 4 SERVINGS

- 6 tablespoons (¾ stick) butter, divided
- 1 large sweet onion, thinly sliced (about 3 cups)
- 1 can (about 14 ounces) chicken broth
- 2 cubes chicken bouillon
- ¼ teaspoon black pepper
- ¼ cup all-purpose flour
- 1½ cups milk
- 1½ cups (6 ounces) shredded Colby-Jack cheese
- Chopped fresh parsley (optional)

1. Melt 2 tablespoons butter in large saucepan or Dutch oven over medium heat. Add onion; cook 10 minutes or until soft and translucent, stirring occasionally.
2. Stir in broth, bouillon and pepper; cook until bouillon is dissolved and mixture is heated through.
3. Meanwhile, melt remaining 4 tablespoons butter in medium saucepan over medium heat. Add flour; cook and stir 1 minute. Gradually whisk in milk until well blended. Cook 10 minutes or until very thick, stirring occasionally.
4. Add milk mixture to broth mixture; cook over medium-low heat 5 to 10 minutes or until thickened, stirring occasionally. Add cheese; cook and stir 5 minutes or until melted and smooth. Garnish with parsley.

Dublin Coddle

MAKES 6 SERVINGS

- 8 ounces Irish bacon*
- 8 pork sausages, preferably Irish bangers
- 3 onions, sliced
- Black pepper
- 2 pounds potatoes, peeled and thickly sliced
- 2 carrots, peeled and cut into 1½-inch pieces
- ¼ cup chopped fresh parsley, plus additional for garnish
- 2 sprigs fresh thyme
- 3 cups chicken broth or water

**Or substitute Canadian bacon or pancetta.*

1. Cook bacon in large saucepan or Dutch oven over medium heat until crisp. Remove to paper towel-lined plate; cut into 1-inch pieces. Drain all but 1 tablespoon drippings from saucepan.
2. Add sausages to Dutch oven; cook about 10 minutes or until browned on all sides. Remove to paper towel-lined plate; cut into 1-inch pieces.
3. Add onions to Dutch oven; cook and stir about 8 minutes or until translucent. Return bacon and sausages to Dutch oven; sprinkle with pepper. Add potatoes, carrots, ¼ cup parsley and thyme; sprinkle generously with pepper. Pour broth over vegetables; bring to a boil.
4. Reduce heat to low; partially cover and cook about 1 hour 20 minutes or until vegetables are tender. Sprinkle with additional parsley, if desired.

Curried Parsnip Soup

MAKES 6 TO 8 SERVINGS

- 3 pounds parsnips, peeled and cut into 2-inch pieces
- 1 tablespoon olive oil
- 2 tablespoons butter
- 1 medium yellow onion, chopped
- 2 stalks celery, diced
- 1 tablespoon salt
- 3 cloves garlic, minced
- 1 to 2 teaspoons curry powder
- ½ teaspoon grated fresh ginger
- ½ teaspoon black pepper
- 8 cups reduced-sodium chicken broth
- Toasted bread slices (optional)
- Chopped fresh chives (optional)

1. Preheat oven to 400ºF. Line large baking sheet with foil.
2. Combine parsnips and oil in large bowl; toss to coat. Spread in single layer on prepared baking sheet. Bake 35 to 45 minutes or until parsnips are tender and lightly browned around edges, stirring once halfway through cooking.
3. Melt butter in large saucepan or Dutch oven over medium heat. Add onion and celery; cook and stir about 8 minutes or until vegetables are tender and onion is translucent. Add salt, garlic, curry powder, ginger and pepper; cook and stir 1 minute. Add parsnips and broth; bring to a boil over medium-high heat. Reduce heat to medium-low; cover and simmer 10 minutes.
4. Working in batches, blend soup in blender or food processor until smooth. (Or use hand-held immersion blender.) Serve with toasted bread, if desired; garnish with chives.

Mushroom Barley Soup

MAKES 6 SERVINGS

- 6 slices bacon
- 1 onion, diced
- 3 stalks celery, sliced
- 2 small carrots, peeled and sliced
- 10 ounces sliced mushrooms
- 1 teaspoon minced garlic
- ⅛ teaspoon red pepper flakes
- 4 cups beef broth
- 1½ cups water
- 1 cup uncooked medium barley, rinsed and drained
- 2 tablespoons balsamic vinegar
- 1 tablespoon Worcestershire sauce
- ½ teaspoon salt
- ¼ teaspoon black pepper

1. Cook bacon in large saucepan or Dutch oven over medium-high heat about 10 minutes or until crisp and browned. Remove to paper towel-lined plate with slotted spoon. Drain off all but 1 tablespoon drippings. Cut bacon into bite-sized pieces.
2. Add onion, celery, carrots and mushrooms to saucepan; cook about 8 minutes until vegetables are crisp-tender, stirring occasionally. Add garlic and red pepper flakes and cook and stir 1 minute.
3. Stir in broth, water, barley, vinegar, Worcestershire sauce, salt and pepper; bring to a boil. Reduce heat to medium-low; stir in bacon. Cover and cook 40 minutes or until barley is tender.

Irish Beef Stew

MAKES 6 SERVINGS

- 2½ tablespoons vegetable oil, divided
- 2 pounds boneless beef chuck roast, cut into 1-inch pieces
- 1½ teaspoons salt, divided
- ¾ teaspoon black pepper, divided
- 1 medium onion, chopped
- 3 medium carrots, cut into 1-inch pieces
- 3 medium parsnips, cut into 1-inch pieces
- 1 package (8 to 10 ounces) cremini mushrooms, quartered
- 2 cloves garlic, minced
- 1 teaspoon dried thyme
- 1 teaspoon dried rosemary
- 2 bay leaves
- 1 can (about 15 ounces) Guinness stout
- 1 can (about 14 ounces) beef broth
- 1 tablespoon Dijon mustard
- 1 tablespoon tomato paste
- 1 tablespoon Worcestershire sauce
- 1 pound small yellow potatoes (about 1¼ inches), halved
- 1 cup frozen pearl onions
- 2 teaspoons water
- 2 teaspoons cornstarch
- Chopped fresh parsley (optional)

1. Heat 2 tablespoons oil in Dutch oven or large saucepan over medium-high heat. Season beef with 1 teaspoon salt and ½ teaspoon pepper. Cook beef in two batches 5 minutes or until browned. Remove to plate.
2. Add remaining ½ tablespoon oil and chopped onion to Dutch oven; cook and stir 3 minutes or until softened. Add carrots, parsnips and mushrooms; cook 8 minutes or until vegetables soften and mushrooms release their liquid, stirring occasionally. Add garlic, thyme, rosemary, bay leaves, remaining ½ teaspoon salt and ¼ teaspoon pepper; cook and stir 2 minutes. Stir in Guinness, broth, mustard, tomato paste and Worcestershire sauce; bring to a boil, scraping up browned bits from bottom of Dutch oven. Return beef to Dutch oven; mix well.
3. Reduce heat to low; cover and cook 1 hour 30 minutes. Stir in potatoes; cover and cook 30 minutes. Stir in pearl onions; cook, uncovered, 30 minutes or until beef and potatoes are fork-tender.
4. Stir water into cornstarch in small bowl until smooth. Add to stew; cook and stir over medium heat 3 minutes or until thickened. Garnish with parsley.

Creamy Fish Chowder

MAKES 4 TO 6 SERVINGS

- 4 ounces bacon, diced
- 1 cup chopped onion
- ½ cup chopped celery
- 2 cups diced peeled russet potatoes
- 2 tablespoons all-purpose flour
- 2 cups water
- 1 teaspoon salt
- 1 bay leaf
- 1 teaspoon dried dill weed
- ½ teaspoon dried thyme
- ½ teaspoon black pepper
- 1 pound cod, haddock or halibut fillets, skinned, boned and cut into 1-inch pieces
- 2 cups milk or half-and-half

1. Cook bacon in large saucepan or Dutch oven over medium-high heat until almost crisp, stirring occasionally. Drain on paper towel-lined plate.
2. Add onion and celery to drippings in saucepan; cook and stir about 5 minutes or until onion is soft. Add potatoes; cook and stir 1 minute. Add flour; cook and stir 1 minute.
3. Stir in water, salt, bay leaf, dill weed, thyme and pepper; bring to a boil over high heat. Reduce heat to low; cover and simmer 25 minutes or until potatoes are fork-tender.
4. Add fish to saucepan; cover and simmer 5 minutes or until fish begins to flake when tested with fork. Remove and discard bay leaf. Stir in bacon. Add milk; cook and stir just until heated through. *Do not boil.*

Brunswick Stew

MAKES 6 TO 8 SERVINGS

- 1 cut-up whole chicken (about 4 pounds)
- 2 quarts water
- 1 stalk celery (including leaves), cut into 2-inch pieces
- 1 onion, quartered
- 2½ teaspoons salt, divided
- 1 clove garlic, halved
- 1 teaspoon whole black peppercorns
- 1 can (about 14 ounces) diced tomatoes
- 2 russet potatoes, peeled and cubed
- 1 onion, thinly sliced
- ¼ cup tomato paste
- 1 teaspoon sugar
- ½ teaspoon dried thyme
- ½ teaspoon ground black pepper
- ⅛ teaspoon garlic powder
- Dash hot pepper sauce
- 1 package (10 ounces) frozen lima beans
- 1 package (10 ounces) frozen corn

1. Combine chicken and water in Dutch oven; bring to a boil over medium-high heat. Skim off foam. Add celery, quartered onion, 2 teaspoons salt, garlic and peppercorns; return to a boil. Reduce heat to medium-low; cover and simmer 2½ to 3 hours.
2. Transfer chicken to cutting board; cool slightly. Remove meat, discarding skin and bones. Cut enough chicken into 1-inch pieces to measure 3 cups. (Reserve remaining chicken for another use.)
3. Strain broth and discard vegetables; skim off fat. Return 1 quart broth to Dutch oven. (Reserve remaining broth for another use.)
4. Add tomatoes, potatoes, sliced onion, tomato paste, sugar, remaining ½ teaspoon salt, thyme, pepper, garlic powder and hot pepper sauce; bring to a boil over medium-high heat. Reduce heat to medium-low; cover and cook 30 minutes. Add beans and corn; cover and cook 5 minutes. Stir in 3 cups chicken; cook 5 minutes or until heated through.

Cream of Broccoli Soup

MAKES 8 SERVINGS

- 1 bunch broccoli (about 1½ pounds), plus additional for garnish
- 3 cups chicken broth
- 1 potato, peeled and chopped
- 1 medium onion, chopped
- 1 carrot, chopped
- 1 stalk celery, chopped
- 1 clove garlic, minced
- ¾ teaspoon salt, divided
- ½ teaspoon dried basil
- 2 tablespoons butter
- 2 tablespoons all-purpose flour
- 1½ cups milk
- 1 cup half-and-half
- ½ cup (2 ounces) shredded Cheddar cheese, plus additional for garnish
- ¼ teaspoon black pepper

1. Trim leaves and ends from broccoli stalks. Peel stalks; cut broccoli into ½-inch pieces.
2. Combine broth, potato, onion, carrot, celery, garlic, ¼ teaspoon salt and basil in medium saucepan; bring to a boil over high heat. Reduce heat to low; cook 10 minutes. Add 1 bunch broccoli to saucepan; cook 10 minutes or until vegetables are fork-tender. Let stand at room temperature 20 to 30 minutes. *Do not drain.*
3. Process vegetable mixture in small batches in food processor or blender until smooth. (Or use hand-held immersion blender.)
4. Melt butter in large saucepan or Dutch oven over medium heat. Stir in flour until smooth. Cook and stir 1 minute. Gradually whisk in milk and half-and-half until well blended. Stir in ½ cup cheese, remaining ½ teaspoon salt and pepper. Add puréed vegetable mixture; cook 3 to 5 minutes or until soup thickens, stirring occasionally. Garnish with additional broccoli and cheese.

Sausage and Bean Stew

MAKES 4 TO 6 SERVINGS

- 2 cups fresh bread crumbs*
- 2 tablespoons olive oil, divided
- 1 pound uncooked pork sausage, cut into 2-inch pieces
- 1 leek, cut in half lengthwise and thinly sliced
- 1 large onion, cut into quarters and cut into ¼-inch slices
- 1 teaspoon salt, divided
- 2 cloves garlic, minced
- ½ teaspoon dried thyme
- ½ teaspoon ground sage
- ¼ teaspoon paprika
- ¼ teaspoon ground allspice
- ¼ teaspoon black pepper
- 1 can (28 ounces) diced tomatoes
- 2 cans (about 15 ounces each) navy or cannellini beans, rinsed and drained
- 2 tablespoons whole grain mustard
- Fresh thyme leaves (optional)

**To make bread crumbs, cut 4 ounces stale baguette or country bread into several pieces; pulse in food processor until coarse crumbs form.*

1. Preheat oven to 350°F. Combine bread crumbs and 1 tablespoon oil in medium bowl; mix well.
2. Heat remaining 1 tablespoon oil in large ovenproof skillet over medium-high heat. Add sausage; cook 8 minutes or until browned, stirring occasionally. (Sausage will not be cooked through.) Remove to plate.
3. Add leek, onion and ½ teaspoon salt to skillet; cook 10 minutes or until vegetables are soft and beginning to brown, stirring occasionally. Add garlic; cook and stir 1 minute. Add dried thyme, sage, paprika, allspice and pepper; cook and stir 1 minute. Add tomatoes; cook 5 minutes, stirring occasionally. Stir in beans, mustard and remaining ½ teaspoon salt; bring to a simmer.
4. Return sausage to skillet, pushing down into bean mixture. Sprinkle with bread crumbs.
5. Bake 25 minutes or until bread crumbs are lightly browned and sausage is cooked through. Garnish with fresh thyme.

Chicken, Barley and Vegetable Soup

MAKES 6 SERVINGS

- 8 ounces boneless skinless chicken breasts, cut into ½-inch pieces
- 8 ounces boneless skinless chicken thighs, cut into ½-inch pieces
- ¾ teaspoon salt
- ¼ teaspoon black pepper
- 1 tablespoon olive oil
- ½ cup uncooked pearl barley
- 4 cans (about 14 ounces each) chicken broth
- 2 cups water
- 1 bay leaf
- 2 cups baby carrots
- 2 cups diced peeled potatoes
- 2 cups sliced mushrooms
- 2 cups frozen peas
- 3 tablespoons sour cream
- 1 tablespoon chopped fresh dill *or* 1 teaspoon dried dill weed

1. Sprinkle chicken with salt and pepper. Heat oil in large saucepan or Dutch oven over medium-high heat. Add chicken; cook without stirring 2 minutes or until golden brown. Turn chicken; cook 2 minutes. Remove to plate.
2. Add barley to saucepan; cook and stir 1 to 2 minutes or until barley begins to brown, adding 1 tablespoon broth, if necessary, to prevent burning. Add remaining broth, water and bay leaf; bring to a boil. Reduce heat to low; cover and simmer 30 minutes.
3. Add chicken, carrots, potatoes and mushrooms to saucepan; cook 10 minutes or until vegetables are tender, stirring occasionally. Add peas; cook 2 minutes. Remove and discard bay leaf.
4. Top soup with sour cream and dill; serve immediately.

HEARTY SANDWICHES

OPEN-FACED LAMB SANDWICHES

MAKES 4 SERVINGS

- 1 tablespoon olive oil
- 1 red onion, diced
- 1 pound ground lamb
- 1¼ teaspoons minced garlic, divided
- 1 tablespoon tomato paste
- 1 teaspoon ground cumin
- ½ teaspoon ground corinader
- 1¼ teaspoons salt, divided
- ¾ cup (6 ounces) plain Greek yogurt
- ¼ cup diced English cucumber
- 2 tablespoons chopped fresh cilantro
- 4 pieces naan bread or pita bread rounds, lightly toasted

1. Heat oil in large skillet over medium heat. Add onion; cook 8 to 10 minutes or until softened. Transfer to small bowl.
2. Cook lamb in same skillet over medium-high heat about 8 minutes or until browned, stirring occasionally. Add 1 teaspoon garlic, tomato paste, cumin, coriander and 1 teaspoon salt; cook 5 minutes, stirring frequently. Add onion; cook and stir 1 minute.
3. Combine yogurt, cucumber, cilantro, remaining ¼ teaspoon garlic and ¼ teaspoon salt in medium bowl; mix well.
4. Divide lamb mixture evenly among naan; top with yogurt mixture. Serve immediately.

Rustic Vegetable Sandwich

MAKES 4 SERVINGS

- 1 pound cremini mushrooms, stemmed and thinly sliced (⅛-inch slices)
- 2 tablespoons olive oil, divided
- ¾ teaspoon salt, divided
- ¼ teaspoon black pepper
- 1 medium zucchini, diced (¼-inch pieces, about 2 cups)
- 3 tablespoons butter, softened
- 8 slices artisan whole grain bread
- ¼ cup prepared pesto
- ¼ cup mayonnaise
- 2 cups packed baby spinach
- 4 slices mozzarella cheese

1. Preheat oven to 350°F. Combine mushrooms, 1 tablespoon oil, ½ teaspoon salt and pepper in medium bowl; toss to coat. Spread in single layer on large baking sheet. Roast 20 minutes or until mushrooms are dark brown and dry, stirring after 10 minutes. Cool on baking sheet.
2. Meanwhile, heat remaining 1 tablespoon oil in large skillet over medium heat. Add zucchini and remaining ¼ teaspoon salt; cook and stir 5 minutes or until zucchini is tender and lightly browned. Remove to medium bowl; wipe out skillet with paper towel.
3. Spread butter on one side of each bread slice. Turn over slices. Spread pesto on 4 bread slices; spread mayonnaise on remaining 4 slices. Top pesto-covered slices evenly with mushrooms; layer with spinach, zucchini and cheese. Top with remaining bread slices, mayonnaise side down.
4. Heat same skillet over medium heat. Add sandwiches; cover and cook 2 minutes per side or until bread is toasted, spinach is slightly wilted and cheese is beginning to melt. Serve immediately.

Classic Patty Melts

MAKES 4 SERVINGS

- **5 tablespoons butter, divided**
- **2 large yellow onions, thinly sliced**
- **¾ teaspoon plus pinch of salt, divided**
- **1 pound ground beef chuck (80% lean)**
- **½ teaspoon garlic powder**
- **½ teaspoon onion powder**
- **¼ teaspoon black pepper**
- **8 slices marble rye bread**
- **½ cup Thousand Island dressing**
- **8 slices (about 1 ounce each) deli American or Swiss cheese**

1. Melt 2 tablespoons butter in large skillet over medium heat. Add onions and pinch of salt; cook 20 minutes or until onions are very soft and golden brown, stirring occasionally. Remove to small bowl; wipe out skillet with paper towel.
2. Combine beef, remaining ¾ teaspoon salt, garlic powder, onion powder and pepper in medium bowl; mix gently. Shape into 4 patties about the size and shape of bread slices and ¼ to ½ inch thick.
3. Melt 1 tablespoon butter in same skillet over medium-high heat. Add patties, two at a time; cook 3 minutes or until bottoms are browned, pressing down gently with spatula to form crust. Turn patties; cook 3 minutes or until browned. Remove patties to plate; wipe out skillet with paper towel.
4. Spread one side of each bread slice with dressing. Top 4 bread slices with 1 cheese slice, patty, caramelized onions, another cheese slice and remaining bread slices.
5. Melt 1 tablespoon butter in same skillet over medium heat. Add 2 sandwiches to skillet; cook 4 minutes or until golden brown, pressing down with spatula to crisp bread. Turn sandwiches; cook 4 minutes or until golden brown and cheese is melted. Repeat with remaining 1 tablespoon butter and sandwiches.

Chicken and Roasted Tomato Sandwiches

MAKES 4 SERVINGS

- 12 ounces plum tomatoes (about 2 large), cut into ⅛-inch slices
- ½ teaspoon coarse salt, divided
- ¼ teaspoon black pepper, divided
- 2 tablespoons olive oil, divided
- 4 boneless skinless chicken breasts (about 4 ounces each)
- 3 tablespoons butter, softened
- ¼ teaspoon garlic powder
- ¼ cup mayonnaise
- 2 tablespoons pesto sauce
- 8 slices sourdough or rustic Italian bread
- 8 slices (about 1 ounce each) provolone cheese
- ½ cup baby spinach

1. Preheat oven to 400°F. Line baking sheet with parchment paper. Arrange tomato slices in single layer on prepared baking sheet. Sprinkle with ¼ teaspoon salt and ⅛ teaspoon pepper; drizzle with 1 tablespoon oil. Roast 25 minutes or until tomatoes are softened and begin to brown around edges.
2. Meanwhile, prepare chicken. If chicken breasts are thicker than ½ inch, pound to ½-inch thickness with meat mallet or rolling pin. Heat remaining 1 tablespoon oil in large skillet over medium-high heat. Season both sides of chicken with remaining ¼ teaspoon salt and ⅛ teaspoon pepper. Add to skillet; cook about 6 minutes per side or until golden brown and no longer pink in center. Remove to plate; let stand 10 minutes before slicing. Cut diagonally into ½-inch slices.
3. Combine butter and garlic powder in small bowl; mix well. Combine mayonnaise and pesto in another small bowl; mix well.
4. Spread one side of each bread slice with garlic butter. For each sandwich, place 1 bread slice, buttered side down, on plate. Spread with generous 1 tablespoon pesto mayonnaise. Layer with 1 cheese slice, 4 to 5 roasted tomato slices, 4 to 6 spinach leaves, 1 sliced chicken breast, another cheese slice and 4 to 6 spinach leaves. Top with second bread slice, buttered side up.
5. Preheat panini press, indoor grill or grill pan. Cook sandwiches until bread is golden brown and cheese is melted.

Bacon-Tomato Grilled Cheese

MAKES 4 SERVINGS

- 8 slices bacon, cut in half
- 4 slices sharp Cheddar cheese
- 4 slices Gouda cheese
- 4 tomato slices
- 8 slices whole wheat or white bread
- 1 tablespoon butter

1. Cook bacon in large skillet over medium-high heat until crisp. Remove from skillet; drain on paper towels. Drain fat from skillet; wipe out skillet with paper towels.
2. Layer 1 Cheddar cheese slice, 1 Gouda cheese slice, 1 tomato slice and 2 bacon slices between 2 bread slices.
3. Melt 1 tablespoon butter in same skillet over medium heat. Add sandwiches; cook 3 to 4 minutes or until bottoms are toasted. Turn sandwiches. Reduce heat to medium-low; cover and cook 3 to 4 minutes or until bottoms are toasted and cheese is melted.

Grilled Reubens with Coleslaw

MAKES 4 SERVINGS

- 2 cups sauerkraut
- ¼ cup (½ stick) butter, softened
- 8 slices marble rye or rye bread
- 12 ounces thinly sliced deli corned beef or pastrami
- ¼ to ½ cup Thousand Island dressing
- 4 slices Swiss cheese
- 2 cups deli coleslaw
- 4 kosher garlic pickle spears

1. Preheat indoor grill or large grill pan. Drain sauerkraut well on paper towel-lined plate.
2. Spread butter evenly over one side of each bread slice. Turn 4 bread slices over; top evenly with corned beef, 1 to 2 tablespoons dressing, sauerkraut and cheese. Top with remaining 4 bread slices, butter side up.
3. Grill sandwiches 4 minutes or until bread is golden brown and cheese begins to melt. Serve with coleslaw and pickles.

TIP

Stack the sandwich ingredients in the order given to prevent sogginess.

Crispy Chicken Sandwich

MAKES 4 SERVINGS

- 2 boneless skinless chicken breasts (6 to 8 ounces each)
- 4 cups cold water
- ¼ cup granulated sugar
- 3 tablespoons plus 1 teaspoon salt, divided
- Vegetable oil for frying
- 1 cup milk
- 2 eggs
- 1½ cups all-purpose flour
- 2 tablespoons powdered sugar
- 1 teaspoon paprika
- 1 teaspoon black pepper
- ¾ teaspoon baking powder
- ¼ teaspoon ground red pepper
- 8 dill pickle slices
- 4 soft hamburger buns, toasted and buttered

1. Pound chicken to ½-inch thickness between two sheets of waxed paper or plastic wrap with rolling pin or meat mallet. Cut each breast in half crosswise to create total of 4 pieces.
2. Combine water, granulated sugar and 3 tablespoons salt in medium bowl; stir until sugar and salt are dissolved. Add chicken to brine; cover and refrigerate 2 to 4 hours. Remove chicken from refrigerator about 30 minutes before cooking.
3. Heat at least 3 inches of oil in large saucepan over medium-high heat to 350°F; adjust heat to maintain temperature. Meanwhile, beat milk and eggs in medium shallow dish until blended. Combine flour, powdered sugar, remaining 1 teaspoon salt, paprika, black pepper, baking powder and red pepper in another shallow dish.
4. Working with one piece at a time, remove chicken from brine and add to milk mixture, turning to coat. Place in flour mixture; turn to coat completely and shake off excess.
5. Lower chicken gently into hot oil; cook 6 to 8 minutes or until no longer pink in center and crust is golden brown and crisp, turning occasionally. Drain on paper towel-lined plate.
6. Place 2 pickle slices on bottom halves of buns; top with chicken and top halves of buns. Serve immediately.

SAVORY LAMB BURGERS

MAKES 4 SERVINGS

- ¼ cup pine nuts
- ¼ cup plain yogurt or sour cream
- 3 cloves garlic, minced, divided
- ¼ teaspoon sugar
- 1 pound ground lamb
- ¼ cup finely chopped yellow onion
- ¾ teaspoon salt
- ¼ teaspoon black pepper
- 4 slices red onion (¼ inch thick)
- 1 tablespoon olive oil
- 8 slices pumpernickel bread
- 4 tomato slices
- 12 thin cucumber slices

1. Cook pine nuts in small skillet over medium heat 1 to 2 minutes or until lightly browned, stirring frequently.
2. Prepare grill for direct cooking over medium-high heat. Oil grid. Whisk yogurt, 1 clove garlic and sugar in small bowl until well blended.
3. Combine lamb, pine nuts, chopped onion, remaining 2 cloves garlic, salt and pepper in large bowl; mix well. Shape mixture into 4 patties about ½ inch thick and 4 inches in diameter. Brush one side of each patty and red onion slice with oil.
4. Place patties and onion on grid, oiled sides down; brush tops with oil. Grill, covered, 4 to 5 minutes per side or until burgers are cooked through (160°F). Grill bread about 1 minute per side during last few minutes of grilling.
5. Serve burgers on bread with red onion, tomato, cucumber and yogurt sauce.

Grilled Portobello Sandwiches

MAKES 4 SERVINGS

- 2 tablespoons extra virgin olive oil
- 1½ tablespoons balsamic vinegar
- 1 tablespoon water
- 1 tablespoon coarse grain Dijon mustard
- 1 teaspoon dried oregano
- 1 clove garlic, minced
- ½ teaspoon black pepper
- ¼ teaspoon salt
- 4 large portobello mushroom caps, wiped with damp towel, gills and stems removed
- 8 slices Italian bread
- 2 tablespoons butter, melted
- 2 to 3 ounces spring greens
- ¼ cup crumbled blue cheese
- 2 to 3 ounces spring greens

1. Whisk oil, vinegar, water, mustard, oregano, garlic, pepper and salt in medium bowl until well blended. Place mushrooms on sheet of foil or large plate. Brush 2 tablespoons dressing over mushrooms; set aside remaining dressing. Let mushrooms stand 30 minutes.
2. Spray grill pan with nonstick cooking spray; heat over medium-high heat. Brush both sides of bread slices lightly with butter. Grill bread 1 minute per side or untli toasted.
3. Grill mushrooms 3 to 4 minutes per side or until tender. Place 1 mushroom on each of 4 bread slices; sprinkle with cheese.
4. Combine greens and reserved dressing; toss gently to coat. Arrange greens over mushrooms; top with remaining bread slices.

Open-Faced Steak and Blue Cheese Sandwiches

MAKES 4 SERVINGS

- 4 boneless beef top loin (strip) or tenderloin steaks, cut ¾-inch thick
- ½ teaspoon plus pinch of salt, divided
- ¼ teaspoon black pepper
- 1½ tablespoons olive oil, divided
- 1 medium onion, cut crosswise into ½-inch slices (keep slices together; do not separate)
- 4 slices ciabatta bread, toasted
- 8 thin slices blue cheese

1. Season steaks with ¼ teaspoon salt and pepper. Heat 1 tablespoon oil in large skillet over medium heat.
2. Add steaks to skillet in single layer (do not crowd). Cook 5 to 6 minutes per side for medium rare (130°F) or until desired doneness. Remove steaks to cutting board; tent with foil and let stand 5 to 10 minutes before slicing.
3. Add remaining ½ tablespoon oil and onion to skillet; cook 3 minutes per side or until edges begin to brown. Add pinch of salt; stir to separate onion into rings.
4. Slice steaks; season with remaining ¼ teaspoon salt. Place 2 slices blue cheese on each bread slice; top with steak and onion. Serve immediately.

FIRESIDE ENTRÉES

FISH AND CHIPS

MAKES 4 SERVINGS

- ¾ cup all-purpose flour
- ½ cup flat beer
- Vegetable oil, for frying
- 4 medium russet potatoes, unpeeled and each cut into 8 wedges
- Salt
- 1 egg, separated
- 1 pound cod fillets (about 6 to 8 small fillets)
- Malt vinegar (optional)
- Lemon wedges (optional)

1. Whisk flour, beer and 2 teaspoons oil in small bowl until well blended. Cover and refrigerate 1 to 2 hours. Line baking sheet with paper towels.
2. Heat 2 inches oil in large heavy skillet over medium heat to 365°F. Add potatoes in batches (do not crowd); cook 4 to 6 minutes or until browned, turning once. Allow temperature of oil to return to 365°F between batches. Drain potatoes on prepared baking sheet; sprinkle lightly with salt. Reserve oil to cook fish.
3. Stir egg yolk into flour mixture. Beat egg white in medium bowl with electric mixer at medium-high speed until soft peaks form. Fold egg white into flour mixture until blended.
4. Return oil to 365°F over medium heat. Dip fish pieces into batter in batches; cook 4 to 6 minutes or until golden brown and fish begins to flake when tested with fork, turning once. Drain on paper towel-lined baking sheet. Serve immediately with potatoes. Sprinkle with vinegar and serve with lemon wedges, if desired.

SHEPHERD'S PIE

MAKES 4 TO 6 SERVINGS

- 3 medium russet potatoes (1½ pounds), peeled and cut into 1-inch pieces
- ½ cup milk
- 5 tablespoons butter, divided
- 1 teaspoon salt, divided
- ½ teaspoon black pepper, divided
- 2 medium onions, chopped
- 2 medium carrots, finely chopped
- ½ teaspoon dried thyme
- 1½ pounds ground lamb or ground beef
- 3 tablespoons tomato paste
- 1 tablespoon Worcestershire sauce
- 1½ cups reduced-sodium beef broth
- ½ cup frozen peas

1. Preheat oven to 350°F. Spray 1½-quart baking dish with nonstick cooking spray.
2. Place potatoes in large saucepan; add water to cover by 2 inches. Bring to a boil over medium-high heat; cook 16 to 18 minutes or until tender. Drain potatoes; return to saucepan.
3. Heat milk in small saucepan over medium-high heat until hot. Add 3 tablespoons butter, ½ teaspoon salt and ¼ teaspoon pepper; stir until butter is melted. Pour milk mixture into saucepan with potatoes; mash until smooth. Set aside.
4. Melt remaining 2 tablespoons butter in large skillet over medium heat. Add onions, carrots and thyme; cook 8 to 10 minutes or until vegetables are softened but not browned, stirring occasionally. Add lamb; cook over medium-high heat 4 minutes or until no longer pink. Drain excess fat. Return skillet to heat; cook 5 minutes or until lamb is lightly browned. Add tomato paste and Worcestershire sauce; cook and stir 1 minute. Stir in broth; bring to a boil and cook about 8 minutes or until almost evaporated. Stir in peas, remaining ½ teaspoon salt and ¼ teaspoon pepper; cook 30 seconds. Pour into prepared baking dish.
5. Spread mashed potatoes in even layer over lamb mixture; use spatula to swirl potatoes or fork to make crosshatch design on top.
6. Bake about 35 minutes or until filling is hot and bubbly and potatoes begin to brown.

Roasted Salmon with Irish Whiskey Sauce

MAKES 4 SERVINGS

- 4 salmon fillets (about 6 ounces each)
- ½ teaspoon salt, divided
- ⅛ teaspoon black pepper
- ⅓ cup Irish whiskey
- ¼ cup finely chopped shallots
- 1 tablespoon white wine vinegar
- ½ cup whipping cream
- 1½ teaspoons Dijon mustard
- 2 tablespoons butter, cut into small pieces
- 2 tablespoons chopped fresh chives

1. Position rack in center of oven. Preheat oven to 425°F. Spray baking sheet with nonstick cooking spray.
2. Sprinkle salmon with ¼ teaspoon salt and pepper; place on prepared baking sheet. Roast 8 to 10 minutes or until fish begins to flake when tested with fork.
3. Meanwhile, combine whiskey, shallots and vinegar in small saucepan; bring to a boil over medium-high heat. Cook about 4 minutes or until liquid almost evaporates and mixture looks like wet sand. Stir in cream and mustard; cook and stir 2 minutes or until slightly thickened. Remove from heat; whisk in butter, chives and remaining ¼ teaspoon salt.
4. Spoon sauce over salmon. Serve immediately.

Corned Beef and Cabbage

MAKES 8 SERVINGS

- 3½ to 4 pounds packaged corned beef brisket
- 3 carrots, cut into 1½-inch pieces
- 1 bunch fresh parsley
- 2 large sprigs fresh thyme
- 1 head green cabbage (about 2 pounds), cut into 8 wedges
- 1½ pounds unpeeled small red potatoes, quartered
- 1 cup sour cream
- 2 tablespoons prepared horseradish
- ½ teaspoon coarse salt
- Chopped fresh parsley (optional)

1. Combine corned beef and carrots in Dutch oven. Tie parsley and thyme together with kitchen string; add to Dutch oven. Add water to cover beef by 1 inch; bring to a boil over high heat. Reduce heat to medium-low; cover and cook about 2½ hours or until beef is almost tender.
2. Add cabbage and potatoes to Dutch oven; cover and cook about 30 minutes or until beef, cabbage and potatoes are tender.
3. Meanwhile, whisk sour cream, horseradish and ½ teaspoon salt in medium bowl until well blended. Refrigerate until ready to serve.
4. Remove herbs from Dutch oven and discard. Remove beef to cutting board; let stand 10 minutes. Slice beef across the grain. Arrange on serving platter with vegetables; season vegetables with additional salt to taste. Sprinkle with chopped parsley, if desired; serve with horseradish sauce.

Irish Stout Chicken

MAKES 4 SERVINGS

- 2 tablespoons vegetable oil
- 1 medium onion, chopped
- 2 cloves garlic, minced
- 1 cut-up whole chicken (3 to 4 pounds)
- 5 carrots, sliced
- 2 parsnips, sliced
- 1 teaspoon dried thyme
- ¾ teaspoon salt
- ½ teaspoon black pepper
- ¾ cup Irish stout
- 8 ounces sliced mushrooms
- ¾ cup frozen peas

1. Heat oil in large skillet over medium heat. Add onion and garlic; cook and stir 3 minutes or until tender. Remove to small bowl.
2. Add chicken to skillet in batches; cook over medium-high heat about 6 minutes per side or until browned. Remove to plate.
3. Add onion mixture, carrots, parsnips, thyme, salt and pepper to skillet. Pour in stout; bring to a boil over high heat, scraping up browned bits from bottom of skillet. Return chicken to skillet. Reduce heat to low; cover and simmer 35 minutes.
4. Add mushrooms and peas to skillet; cover and cook 10 minutes. Uncover; cook over medium heat 10 minutes or until sauce is slightly thickened and chicken is cooked through (165°F).

PORK TENDERLOIN WITH CABBAGE AND LEEKS

MAKES 4 SERVINGS

- ¼ cup olive oil, plus additional for pan
- 1 teaspoon salt
- ¾ teaspoon garlic powder
- ½ teaspoon dried thyme
- ½ teaspoon black pepper
- 1 pork tenderloin (about 1¼ pounds)
- ½ medium savoy cabbage, cored and cut into ¼-inch slices (about 6 cups)
- 1 small leek, cut in half lengthwise and cut crosswise into ¼-inch slices
- 1 to 2 teaspoons cider vinegar

1. Preheat oven to 450°F. Brush baking sheet with oil.
2. Combine salt, garlic powder, thyme and pepper in small bowl; mix well. Stir in ¼ cup oil until well blended. Brush pork with about 1 tablespoon oil mixture, turning to coat all sides.
3. Combine cabbage and leek in large bowl. Drizzle with remaining oil mixture; toss to coat. Spread on prepared baking sheet; top with pork.
4. Bake 25 minutes or until pork is 145°F, stirring cabbage mixture halfway through cooking. Remove pork to cutting board; tent with foil to keep warm. Let stand 10 minutes before slicing. Add vinegar to cabbage mixture; stir to blend.

TIPS

If you can't find savoy cabbage, you can substitute regular green cabbage but it may take slightly longer to cook. If the cabbage is not crisp-tender when the pork is done, return the vegetables to the oven for 10 minutes or until crisp-tender.

LAMB AND MINT HAND PIES

MAKES 4 MAIN-DISH OR 8 APPETIZER SERVINGS

- 2 cups plus 1 tablespoon all-purpose flour, divided
- 1 teaspoon salt, divided
- 10 tablespoons cold butter, cut into small pieces
- 7 to 8 tablespoons ice water
- 1 pound ground lamb
- 1 small onion, finely chopped
- 1 carrot, finely chopped
- ½ cup reduced-sodium beef broth
- 1 teaspoon Dijon mustard
- ¼ teaspoon black pepper
- 1 tablespoon chopped fresh mint
- ½ cup (2 ounces) shredded Irish Cheddar cheese
- 1 egg, lightly beaten

1. Combine 2 cups flour and ½ teaspoon salt in medium bowl. Cut in butter with pastry blender until mixture resembles coarse crumbs. Add water, 1 tablespoon at a time, stirring with fork until loose dough forms. Knead dough in bowl 1 to 2 times until it comes together. Divide dough into four pieces; press each into 4-inch disc. Wrap dough in plastic wrap; freeze 15 minutes.
2. Meanwhile, prepare filling. Heat large skillet over medium-high heat. Add lamb; cook 7 to 8 minutes or until lightly browned, stirring occasionally. Remove lamb to bowl; drain fat. Add onion and carrot to skillet; cook 3 minutes or until vegetables begin to soften, stirring occasionally. Stir in lamb; cook 1 minute. Add remaining 1 tablespoon flour; cook and stir 1 minute. Add broth, mustard, remaining ½ teaspoon salt and pepper; cook over medium heat 2 minutes or until thickened. Remove from heat; stir in mint. Cool 10 minutes. Stir in cheese.
3. Position rack in center of oven. Preheat oven to 400°F. Line large baking sheet with parchment paper or spray with nonstick cooking spray.
4. Working with one disc at a time, roll out dough into 9-inch circle on lightly floured surface. Cut out four circles with 4-inch round cookie cutter (16 circles total). Place eight circles on prepared baking sheet. Spread one eighth of lamb filling over each circle, leaving ½-inch border around edge. Top with remaining dough circles, pressing edges to seal. Press edges again with tines of fork. Brush tops with egg; cut 1-inch slit in top of each pie with tip of knife.
5. Bake 28 to 30 minutes or until golden brown. Serve hot or at room temperature.

TIP

Pies can be made 1 day ahead and refrigerated. Reheat before serving.

Sirloin with Mushrooms and Whiskey Cream Sauce

MAKES 4 SERVINGS

- 2 tablespoons butter
- 1 tablespoon vegetable oil
- 8 ounces cremini mushrooms, sliced
- 1½ pounds sirloin steak
- ½ teaspoon salt
- ¼ teaspoon black pepper
- ½ cup Irish whiskey
- ½ cup whipping cream
- ½ cup reduced-sodium beef broth
- Chopped fresh chives

1. Heat butter and oil in large skillet over medium-high heat. Add mushrooms; cook 8 minutes or until liquid evaporates, stirring occasionally. Remove to bowl.
2. Sprinkle both sides of steak with salt and pepper. Add to skillet; cook about 3 minutes per side or until desired doneness. Remove to plate; tent with foil to keep warm.
3. Add whiskey to skillet; cook 2 minutes, scraping up browned bits from bottom of skillet. Add cream and broth; cook and stir 3 minutes. Stir in any accumulated juices from steak.
4. Return mushrooms to skillet; cook and stir 2 minutes or until sauce thickens. Pour sauce over steak; sprinkle with chives.

BANGERS AND MASH

MAKES 4 TO 6 SERVINGS

- 2 pounds bangers or fresh mild pork sausages
- 2 tablespoons vegetable oil, divided
- 2¼ pounds unpeeled Yukon Gold potatoes, cut into 1-inch pieces
- ¾ cup milk, heated
- 3 tablespoons butter, melted
- 1½ teaspoons coarse salt
- 2 small yellow onions, halved and thinly sliced (about 2 cups)
- 1 tablespoon butter
- 1 tablespoon all-purpose flour
- ¼ cup dry red wine
- 1¼ cups reduced-sodium beef broth
- Black pepper

1. Preheat oven to 400°F. Line baking sheet with foil. Combine sausages and 1 tablespoon oil in large bowl; toss to coat. Place on prepared baking sheet; roast about 20 minutes or until sausages are cooked through and golden brown, turning once halfway through cooking.
2. Meanwhile, place potatoes in large saucepan; add water to cover by 2 inches. Bring to a boil over high heat. Reduce heat to medium-low; cook about 12 minutes or until tender.
3. Drain potatoes well; press through ricer or mash with potato masher in medium bowl. Stir in warm milk, 3 tablespoons melted butter and salt until well blended. Set aside and keep warm.
4. Heat remaining 1 tablespoon oil in medium saucepan over medium heat. Add onions; cover and cook 10 minutes, stirring occasionally. Stir in ½ cup water; cover and cook 10 to 15 minutes or until golden brown, stirring occasionally.
5. Add 1 tablespoon butter to onions; cook and stir until melted. Add flour; cook and stir 1 minute. Add wine; cook about 30 seconds or until almost evaporated. Add broth; cook over medium-high heat about 5 minutes or until thickened, stirring occasionally. Season with additional salt and pepper. Serve gravy with bangers and mashed potatoes.

ROASTED DIJON LAMB AND VEGETABLES

MAKES 8 SERVINGS

- 20 cloves garlic, peeled (about 2 medium heads)
- ¼ cup Dijon mustard
- 2 tablespoons water
- 2 tablespoons fresh rosemary leaves
- 1 tablespoon fresh thyme
- 1¼ teaspoons salt, divided
- 1 teaspoon black pepper
- 4½ pounds boneless leg of lamb,* trimmed
- 1 pound parsnips, cut diagonally into ½-inch pieces
- 1 pound carrots, cut diagonally into ½-inch pieces
- 2 large onions, cut into ½-inch wedges
- 2 tablespoons olive oil

***If unavailable, substitute packaged marinated lamb; rinse well and pat dry.**

1. Combine garlic, mustard, water, rosemary, thyme, ¾ teaspoon salt and pepper in food processor; process until smooth. Place lamb in large bowl or baking pan. Spoon mixture over top and sides of lamb. Cover and refrigerate at least 8 hours.
2. Preheat oven to 500°F. Line broiler pan with foil; top with broiler rack. Spray rack with nonstick cooking spray. Combine parsnips, carrots, onions, oil and ¼ teaspoon salt in large bowl; toss to coat. Spread vegetables evenly on broiler rack; top with lamb.
3. Roast 15 minutes. *Reduce oven temperature to 325°F.* Roast 1 hour 20 minutes or until lamb is 140°F for medium or until desired doneness.
4. Remove lamb to cutting board; tent with foil and let stand 10 minutes before slicing. Stir vegetables; continue roasting 10 minutes.
5. Season vegetables with remaining ¼ teaspoon salt. Slice lamb; serve with vegetables.

TROUT WITH PINE NUT BUTTER

MAKES 4 SERVINGS

- 4 whole trout (about 8 ounces each), cleaned
- ¼ cup olive oil
- ¼ cup dry white wine
- 2 tablespoons minced chives
- 2 tablespoons chopped fresh parsley
- ½ teaspoon salt
- ⅛ teaspoon black pepper
- ¼ cup (½ stick) butter, softened
- ¼ cup pine nuts, finely chopped
- 1 lemon, cut into wedges (optional)

1. Place trout in large resealable food storage bag. Whisk oil, wine, chives, parsley, salt and pepper in small bowl until well blended. Pour over fish; seal bag and turn to coat. Marinate at room temperature 30 minutes or refrigerate up to 2 hours, turning occasionally.
2. Combine butter and pine nuts in small bowl; mix well. Cover and let stand at room temperature until ready to use.
3. Preheat broiler; spray broiler pan with nonstick cooking spray. Remove fish from marinade; reserve marinade. Place fish on prepared pan.
4. Broil 4 to 6 inches from heat 4 minutes. Turn fish; brush with reserved marinade. Broil 4 to 6 minutes or until fish is opaque and begins to flake when tested with fork.
5. Transfer fish to serving plates. Top with butter mixture; garnish with lemon wedges.

SHEPHERD'S PIE STUFFED POTATOES

MAKES 4 SERVINGS

- 4 large russet potatoes (about 12 ounces each)
- 1½ tablespoons olive oil, divided
- 1 small onion, chopped
- 1 carrot, chopped
- 1 clove garlic, minced
- 12 ounces ground beef chuck
- 2 tablespoons tomato paste
- 1 tablespoon Worcestershire sauce
- ½ teaspoon ground thyme
- 1 teaspoon salt, divided
- ½ teaspoon black pepper, divided
- ½ cup water
- ½ cup thawed frozen peas
- 3 to 4 tablespoons milk
- 2 tablespoons butter

1. Preheat oven to 400°F. Scrub potatoes; prick all over with fork. Brush with 1 tablespoon oil; place on baking sheet. Bake about 1 hour or until fork-tender.
2. Meanwhile, heat remaining ½ tablespoon oil in large skillet over medium-high heat. Add onion and carrot; cook about 8 minutes or until vegetables are soft and beginning to brown, stirring occasionally. Add garlic; cook and stir 1 minute. Add beef; cook about 5 minutes or until no longer pink, stirring to break up meat. Add tomato paste, Worcestershire sauce, thyme, ½ teaspoon salt and ¼ teaspoon pepper; cook and stir 2 minutes. Stir in water. Reduce heat to medium-low; cook about 10 minutes or until mixture thickens slightly, stirring occasionally. Stir in peas; cook 1 minute.
3. Cut ½-inch slice from top of each baked potato. Scoop out flesh into medium bowl, leaving ¼-inch shells. Add 3 tablespoons milk, butter, remaining ½ teaspoon salt and ¼ teaspoon pepper to potatoes in bowl; mash until smooth, adding additional milk if necessary. Return potato shells to baking sheet.
4. Divide beef mixture evenly among potato shells. Pipe or spread mashed potato mixture over beef mixture. (There may be extra mashed potatoes; serve on the side or reserve for another use.)
5. Bake about 20 minutes or until tops of potatoes begin to brown.

Herbed Pork with Potatoes and Green Beans

MAKES 4 SERVINGS

- 2 tablespoons chopped fresh thyme
- 2 tablespoons chopped fresh rosemary
- 2 cloves garlic, minced
- 2 teaspoons salt
- ¾ teaspoon black pepper
- ¼ cup olive oil
- 1½ pounds fingerling potatoes (about 18 potatoes), cut in half lengthwise
- 1 pound fresh green beans
- 2 pork tenderloins (about 12 ounces each)

1. Preheat oven to 450°F. Combine thyme, rosemary, garlic, salt and pepper in small bowl. Stir in oil until well blended.
2. Place potatoes in medium bowl. Drizzle with one third of oil mixture, toss to coat. Arrange potatoes, cut sides down, in rows covering two thirds of large baking sheet. (Potatoes should be in single layer; do not overlap.) Leave remaining one third of baking sheet empty.
3. Roast potatoes 10 minutes while preparing beans and pork. Trim beans; place in same bowl used for potatoes. Drizzle with one third of oil mixture; toss to coat.
4. When potatoes have roasted 10 minutes, remove baking sheet from oven. Arrange beans on empty one third of baking sheet. Brush all sides of pork with remaining oil mixture; place on top of green beans.
5. Roast 20 to 25 minutes or until pork is 145°F. Remove pork to large cutting board; tent with foil and let stand 10 minutes. Stir vegetables; return to oven. Roast 10 minutes or until potatoes are dark golden brown and beans are charred in spots. Slice pork; serve with vegetables.

LAMB SHANKS BRAISED IN STOUT

MAKES 4 SERVINGS

- 4 lamb shanks (about 1 pound each)*
- ¼ cup all-purpose flour
- ¼ cup vegetable oil, plus additional as needed
- 1 large onion, chopped (about 2 cups)
- 4 cloves garlic, minced
- Salt and black pepper
- 3 sprigs fresh rosemary
- 3 sprigs fresh thyme
- 1 bottle (about 11 ounces) Irish stout
- 2 to 3 cups reduced-sodium chicken broth
- Smashed Potatoes (recipe follows)
- 1 tablespoon chopped fresh mint

**For a more attractive presentation, ask butcher to "french" chops by removing flesh from last inch of bone end.*

1. Preheat oven to 325°F. Trim excess fat from lamb shanks. (Do not remove all fat or lamb will fall apart while cooking.) Dust lamb with flour. Heat ¼ cup oil in Dutch oven over medium-high heat. Add lamb in batches; cook until browned on all sides. Remove to plate.
2. Add oil to Dutch oven, if necessary, to make about 2 tablespoons. Add onion; cook and stir 2 minutes. Add garlic; cook and stir 2 minutes. Return lamb to Dutch oven; sprinkle generously with salt and pepper. Tuck rosemary and thyme sprigs around lamb. Add stout to pan; pour in broth to almost cover lamb.
3. Cover and bake 2 hours or until lamb is very tender and almost falling off bones. Meanwhile, prepare Smashed Potatoes.
4. Remove lamb to platter; tent with foil to keep warm. Skim fat from juices in Dutch oven; boil over medium-high heat until reduced by half. Strain sauce. Serve lamb and sauce over potatoes; sprinkle with mint.

SMASHED POTATOES

Place 1½ to 2 pounds unpeeled small Yukon gold or white potatoes in large saucepan; add cold water to cover by 2 inches. Bring to a boil over high heat. Reduce heat to medium-low; cook about 20 minutes or until fork-tender. Drain potatoes; return to saucepan and stir in 1 tablespoon butter until melted. Partially smash potatoes with fork or potato masher. Season with salt and black pepper.

VEGETABLES & SIDES

Potato and Corned Beef Cakes

MAKES 10 CAKES

- 2 pounds russet potatoes, divided
- 2 teaspoons salt, divided
- 6 tablespoons all-purpose flour
- ¼ cup whole milk
- 1 egg, beaten
- ½ teaspoon black pepper
- 1 cup chopped corned beef (leftover or deli corned beef, about ⅓ pound), cut into ¼-inch pieces
- 1 tablespoon butter
- 1 tablespoon olive oil
- Chopped fresh parsley (optional)

1. Peel half of potatoes; cut into 1-inch pieces. Place in medium saucepan; add 1 teaspoon salt and water to cover by 2 inches. Bring to a simmer over medium heat; cook 15 minutes or until tender. Drain and rice potatoes into medium bowl.
2. Peel remaining half of potatoes; grate with box grater. Squeeze out and discard liquid. Add grated potatoes to riced potatoes in bowl; stir in flour, milk, egg, remaining 1 teaspoon salt and pepper. Stir in corned beef until well blended.
3. Heat butter and oil in large skillet (preferably cast iron) over medium heat. Shape ⅓ cupfuls of potato mixture into patties; cook in batches 3 to 4 minutes per side or until golden brown. (Do not crowd patties in skillet.) Sprinkle with parsley, if desired.

Red Cabbage with Bacon and Mushrooms

MAKES 6 SERVINGS

- 5 slices thick-cut bacon, chopped (about 8 ounces)
- 1 onion, chopped
- 1 package (8 ounces) cremini mushrooms, chopped (½-inch pieces)
- ¾ teaspoon dried thyme
- ½ medium red cabbage, cut into wedges, cored and cut crosswise into ¼-inch slices (about 7 cups)
- ¼ teaspoon salt
- ¼ teaspoon black pepper
- ⅔ cup chicken broth
- 3 tablespoons cider vinegar
- ¼ cup chopped walnuts, toasted*
- 3 tablespoons chopped fresh parsley

***To toast walnuts, cook in small skillet over medium heat 3 to 4 minutes or until lightly browned, stirring frequently.**

1. Cook bacon in large saucepan or deep skillet over medium-high heat until crisp. Remove to paper towel-lined plate. Drain all but 1 tablespoon drippings from saucepan.
2. Add onion to saucepan; cook and stir 5 minutes or until softened. Add mushrooms and thyme; cook 6 minutes or until mushrooms begin to brown, stirring occasionally. Add cabbage, ¼ teaspoon salt and ¼ teaspoon pepper; cook 7 minutes or until cabbage is wilted.
3. Stir in broth, vinegar and half of bacon; bring to a boil. Reduce heat to low; cook, uncovered, 15 to 20 minutes or until cabbage is tender.
4. Stir in walnuts and parsley; season with additional salt and pepper, if necessary. Sprinkle with remaining bacon.

Brussels Sprouts with Honey Butter

MAKES 4 SERVINGS

- 6 slices thick-cut bacon, cut into ½-inch pieces
- 1½ pounds Brussels sprouts (about 24 medium), halved
- ¼ teaspoon salt
- ¼ teaspoon black pepper
- 2 tablespoons butter, softened
- 2 tablespoons honey

1. Preheat oven to 375°F. Cook bacon in medium skillet until almost crisp. Drain on paper towel-lined plate. Reserve 1 tablespoon drippings.
2. Place Brussels sprouts on large baking sheet. Drizzle with reserved bacon drippings and sprinkle with salt and pepper; toss to coat. Arrange Brussels sprouts cut sides down in single layer on baking sheet.
3. Roast 30 minutes or until Brussels sprouts are browned, stirring once.
4. Whisk butter and honey in medium bowl until well blended. Add warm Brussels sprouts; stir until completely coated. Stir in bacon; season with additional salt and pepper.

Creamy Slab Potatoes

MAKES 4 SERVINGS

- ¼ cup (½ stick) butter, melted
- 1 teaspoon salt
- ½ teaspoon dried rosemary
- ½ teaspoon dried thyme
- ¼ teaspoon black pepper
- 2½ pounds Yukon Gold potatoes (6 to 8 potatoes), peeled and cut crosswise into 1-inch slices
- 1 cup water
- 3 cloves garlic, smashed

1. Preheat oven to 500°F.
2. Combine butter, salt, rosemary, thyme and pepper in 13×9-inch baking pan (do not use glass); mix well. Add potatoes; toss to coat. Spread in single layer.
3. Bake 15 minutes. Turn potatoes; bake 15 minutes. Add water and garlic to pan; bake 15 minutes. Remove to platter; pour any remaining liquid in pan over potatoes.

Rosemary Bread

MAKES 2 SMALL LOAVES

- 1 tablespoon sugar
- 1 package (¼ ounce) active dry yeast
- 1 cup warm water (110°F)
- 2 tablespoons finely chopped fresh rosemary, divided
- 2½ to 2¾ cups all-purpose flour, divided
- 2 tablespoons olive oil
- 1⅛ teaspoons salt, divided
- 2 tablespoons butter, melted, divided
- ⅛ teaspoon black pepper

1. Dissolve sugar and yeast in warm water in large bowl of stand mixer; let stand 5 minutes or until bubbly. Reserve 1 teaspoon rosemary for topping; set aside.
2. Add 2½ cups flour, oil, 1 teaspoon salt and remaining rosemary to yeast mixture; stir until blended. Mix with dough hook at medium speed 5 minutes. Add additional flour, 1 tablespoon at a time, if dough is too sticky or does not clean side of bowl. Shape dough into a ball. Place in large greased bowl; turn to grease top. Cover and let rise in warm place about 1 hour or until doubled in size.
3. Line baking sheet with parchment paper. Gently punch down dough; divide in half. Shape each half into a ball; place on prepared baking sheet. Brush with 1 tablespoon butter; sprinkle with reserved rosemary, remaining ⅛ teaspoon salt and pepper. Cover loosely with plastic wrap and let rise in warm place about 45 minutes or until almost doubled in size. Preheat oven to 400°F.
4. Bake loaves 20 minutes or until golden brown. Brush with remaining 1 tablespoon melted butter. (Brush lightly to avoid removing seasoning from tops of loaves.) Remove to wire rack to cool completely.

OVEN-ROASTED VEGETABLES

MAKES 4 SERVINGS

- 8 ounces cremini mushrooms, halved or quartered if large
- 8 ounces Brussels sprouts, trimmed and quartered
- 2 carrots, cut into ½-inch pieces
- 2 parsnips, cut into ½-inch pieces
- 1 medium red onion, diced
- 1 medium zucchini, diced
- 1 red bell pepper, diced
- 2 tablespoons olive oil
- 1 teaspoon salt
- ¼ teaspoon black pepper
- 1 tablespoon butter, melted
- 1 green onion, finely chopped
- 1 tablespoon chopped fresh basil
- 1 clove garlic, minced
- 1 tablespoon balsamic glaze
- 1 tablespoon grated Parmesan cheese

1. Preheat oven to 450°F. Combine mushrooms, Brussels sprouts, carrots, parsnips, red onion, zucchini, and bell pepper in large bowl. Drizzle with oil; season with salt and pepper and toss to coat. Spread in single layer on baking sheet.
2. Roast 15 minutes. Stir vegetables; roast 5 to 10 minutes or until crisp-tender. *Turn oven to broil.*
3. Drizzle vegetables with butter. Sprinkle with green onion, basil and garlic; stir to coat. Spread vegetables in single layer.
4. Broil 3 to 5 minutes or until edges of vegetables begin to brown. Drizzle with balsamic glaze; sprinkle with cheese.

Individual Potato Gratins

MAKES 12 GRATINS (6 SERVINGS)

- 1 cup whipping cream
- 1 tablespoon chopped fresh thyme
- 1 clove garlic, minced
- 1 teaspoon salt
- ⅛ teaspoon black pepper
- 2 pounds russet potatoes
- ¼ cup grated Parmesan cheese
- 1 cup (4 ounces) grated Gruyère cheese

1. Preheat oven to 375°F. Spray 12 standard (2½-cup) muffin cups with nonstick cooking spray.
2. Pour cream into small microwavable bowl or glass measuring cup. Microwave on HIGH 1 minute or just until cream begins to bubble around edges. Stir in thyme, garlic, salt and pepper until blended; let stand while preparing potatoes.
3. Peel potatoes and cut crosswise into ⅛-inch slices. Layer potato slices in prepared muffin cups, filling half full. Sprinkle with Parmesan; layer remaining potato slices over Parmesan. Pour cream mixture over potatoes; press potato stacks down firmly. Cover pan loosely with foil; place on baking sheet.
4. Bake 30 minutes. Remove pan from oven; sprinkle potatoes with Gruyère. Bake, uncovered, 30 minutes or until potatoes are tender and golden brown. (A paring knife inserted into potatoes should go in easily when potatoes are tender.) Let stand 5 minutes. Use small spatula or knife to loosen edges and bottoms of gratins; remove to plate. Serve warm.

TIP

Gratins can be made ahead, refrigerated and reheated for 10 to 15 minutes in a 350°F oven.

Brown Bread

MAKES 2 LOAVES

- 1 package (¼ ounce) active dry yeast
- 1¼ cups warm water (105° to 110°F)
- ¼ cup packed brown sugar
- ¼ cup molasses
- 2 tablespoons vegetable oil
- 1½ tablespoons unsweetened cocoa powder
- 1¼ teaspoons salt
- 2 cups all-purpose flour
- 1½ cups whole wheat flour
- 2 teaspoons old-fashioned oats

1. Dissolve yeast in warm water in large bowl of stand mixer; let stand 5 minutes or until bubbly. Stir in brown sugar, molasses, oil, cocoa and salt until well blended. Add all-purpose flour and whole wheat flour; stir until rough dough forms.
2. Mix with dough hook at low speed 5 minutes (dough will be slightly sticky but should clean side of bowl). Shape dough into a ball. Place in large greased bowl; turn to grease top. Cover and let rise in warm place 1½ to 2 hours or until doubled in size.
3. Line baking sheet with parchment paper. Turn out dough onto very lightly floured surface; divide in half. Shape each half into 12×3-inch loaf; place loaves on prepared baking sheet. Moisten top and sides of loaves with dampened hands; sprinkle 1 teaspoon oats over top and sides of each loaf, pressing lightly to adhere. Loosely cover loaves and let rise in warm place 1 hour or until very puffy and almost doubled in height.
4. Preheat oven to 350°F. Bake 25 minutes or until bottoms of loaves are lightly browned and instant-read thermometer inserted into center of loaves registers 190°F. Remove to wire racks to cool. Serve warm or cool completely.

Green Beans and Mushrooms

MAKES 4 TO 6 SERVINGS

- 1½ tablespoons olive oil, divided
- 1 small onion, thinly sliced
- 8 ounces sliced mushrooms
- ¾ teaspoon salt, divided
- 1 pound fresh green beans, trimmed
- 1 teaspoon minced garlic
- ¼ teaspoon black pepper

1. Heat 1 tablespoon oil in large skillet over medium-high heat. Add onion; cook and stir 2 minutes or until beginning to soften. Add mushrooms; cook about 5 minutes or until mushrooms release their liquid and begin to brown, stirring occasionally.
2. Add beans, remaining ½ tablespoon oil and ½ teaspoon salt to skillet; cook and stir 3 minutes. Reduce heat to medium-low; cover and cook 10 to 12 minutes or until beans are crisp-tender.
3. Stir in garlic, remaining ¼ teaspoon salt and pepper; cook and stir 2 minutes.

Smashed Potatoes

MAKES 4 SERVINGS

- 4 medium russet potatoes (about 1½ pounds), peeled and cut into ¼-inch pieces
- ⅓ cup milk
- 2 tablespoons sour cream
- 1 tablespoon minced onion
- ½ teaspoon salt
- ¼ teaspoon black pepper
- ⅛ teaspoon garlic powder (optional)
- Chopped fresh chives or French fried onions (optional)

1. Bring large saucepan of lightly salted water to a boil over medium-high heat. Add potatoes; cook 15 to 20 minutes or until fork-tender. Drain and return to saucepan.
2. Slightly mash potatoes. Stir in milk, sour cream, minced onion, salt, pepper and garlic powder, if desired; mash until desired texture is reached, leaving potatoes chunky. Cook 5 minutes over low heat or until heated through, stirring occasionally. Garnish with chives.

Roasted Garlic and Stout Mac and Cheese

MAKES 8 TO 10 SERVINGS

- 6 tablespoons (¾ stick) butter, divided, plus additional for baking dish
- 1 head garlic
- 1 tablespoon olive oil
- 1 cup panko bread crumbs
- 1¼ teaspoons salt, divided
- 1 package (16 ounces) uncooked cellentani pasta*
- ¼ cup all-purpose flour
- ½ teaspoon black pepper
- 2 cups whole milk
- ¾ cup Irish stout
- 2 cups (8 ounces) shredded sharp Cheddar cheese
- 2 cups (8 ounces) shredded Dubliner or white Cheddar cheese

***Or substitute elbow macaroni, penne or other favorite pasta shape.*

1. Preheat oven to 375°F. Butter 4-quart shallow baking dish.
2. Place garlic on 10-inch piece of foil; drizzle with oil and crimp shut. Place on small baking sheet; bake 30 to 40 minutes or until tender. Cool 15 minutes; squeeze cloves into small bowl. Mash into smooth paste.*
3. Melt 2 tablespoons butter in small saucepan over medium heat. Stir in panko and ¼ teaspoon salt until well blended.
4. Cook pasta in large saucepan of salted boiling water according to package directions for al dente. Drain and return to saucepan; keep warm.
5. Meanwhile, melt remaining 4 tablespoons butter in large saucepan over medium heat. Add flour; cook and stir until lightly browned. Stir in roasted garlic, remaining 1 teaspoon salt and pepper. Slowly whisk in milk and stout; cook until thickened, whisking constantly. Remove from heat; whisk in cheeses, ½ cup at a time, until melted. Combine cheese mixture and pasta in large bowl; stir to coat. Spoon into prepared baking dish; sprinkle with panko mixture.
6. Bake 40 minutes or until bubbly and topping is golden brown. Let stand 10 minutes before serving.

**Garlic can be roasted ahead of time; store in refrigerator until ready to use.*

STOVIES WITH BACON

MAKES 4 SERVINGS

- 3 medium russet potatoes (about 1½ pounds), peeled
- 6 slices bacon
- 2 large onions, halved vertically and sliced
- 4 teaspoons butter
- ½ teaspoon salt
- ⅛ teaspoon black pepper
- ⅓ cup water

1. Place potatoes in large saucepan; add cold water to cover by 2 inches. Bring to a boil over medium-high heat; cook 15 minutes or until potatoes are partially cooked. Drain potatoes; let stand until cool enough to handle. Cut potatoes into ½-inch-thick slices.
2. Cook bacon in large skillet over medium-high heat 6 to 7 minutes or until crisp. Drain on paper towel-lined plate. Chop bacon; set aside.
3. Drain all but 2 tablespoons drippings from skillet; heat over medium heat. Add onions; cook 8 to 10 minutes or until softened but not browned, stirring occasionally. Remove to small bowl.
4. Add butter to same skillet; heat over medium heat until melted. Add potatoes; sprinkle with salt and pepper. Top with onions and pour in ⅓ cup water; cover and cook 5 minutes. Stir in bacon; cook, uncovered, 10 to 12 minutes or until potatoes are tender and browned, stirring occasionally.

Brown Soda Bread

MAKES 6 TO 8 SERVINGS

- 2 cups all-purpose flour
- 1 cup whole wheat flour
- 1 teaspoon baking soda
- ½ teaspoon salt
- ½ teaspoon ground ginger
- 1¼ to 1½ cups buttermilk
- 3 tablespoons dark molasses (preferably blackstrap)

1. Preheat oven to 375°F. Line baking sheet with parchment paper.
2. Combine 2 cups all-purpose flour, whole wheat flour, baking soda, salt and ginger in large bowl; mix well. Whisk 1¼ cups buttermilk and molasses in medium bowl until well blended. Stir into flour mixture. Add additional buttermilk, 1 tablespoon at a time, if needed to make dry, rough dough.
3. Turn out dough onto floured surface; knead 8 to 10 times or just until smooth. (Do not overknead.) Shape dough into round loaf about 1½ inches thick. Place on prepared baking sheet.
4. Use floured knife to cut halfway through dough, scoring into quarters. Sprinkle top of dough with additional all-purpose flour, if desired.
5. Bake about 35 minutes or until bread sounds hollow when tapped. Remove to wire rack to cool slightly. Serve warm.

Braised Leeks

MAKES 4 SERVINGS

- **3 to 4 large leeks (1½ to 2 pounds)**
- **¼ cup (½ stick) butter**
- **¼ teaspoon salt**
- **¼ teaspoon black pepper**
- **¼ cup dry white wine**
- **¼ cup reduced-sodium chicken or vegetable broth**
- **3 to 4 sprigs fresh parsley**

1. Trim green stem ends of leeks; remove any damaged outer leaves. Slice leeks lengthwise up to, but not through, root ends to hold leeks together. Rinse leeks in cold water, separating layers to remove embedded dirt. Cut leeks crosswise into 3-inch lengths; cut off and discard root ends.
2. Melt butter in skillet large enough to hold leeks in single layer. Arrange leeks in skillet in crowded layer, keeping pieces together as much as possible. Cook over medium-high heat about 8 minutes or until leeks begin to color and soften, turning with tongs once or twice. Sprinkle with salt and pepper.
3. Add wine, broth and parsley; bring to a simmer. Cover and cook over low heat 20 minutes or until leeks are very tender. Remove parsley sprigs.

TIP

Leeks often contain a lot of embedded dirt between their layers, so they need to be washed thoroughly. It's easiest to slice up to—but not through—the root ends before slicing or chopping so the leeks hold together while washing them.

SERVING SUGGESTION

Top the braised leeks with toasted bread crumbs, grated cheese or crisp crumbled bacon for an extra rich side dish.

DESSERTS & DRINKS

Irish Bread Pudding

MAKES 6 TO 8 SERVINGS

- 14 slices day-old, firm-textured white bread (about 12 ounces), crusts trimmed
- 1½ cups milk
- ⅓ cup butter, softened
- ⅓ cup packed brown sugar
- 1 teaspoon ground cinnamon
- ¼ teaspoon ground nutmeg, plus additional for garnish
- ¼ teaspoon ground cloves
- 1 medium apple, peeled and chopped
- 1 package (6 ounces) mixed dried fruit, chopped
- 1 egg
- ⅓ cup chopped nuts
- Sweetened whipped cream (optional)

1. Tear bread into pieces; place in large bowl. Pour milk over bread; let soak 30 minutes.
2. Preheat oven to 350°F. Spray 9×5-inch loaf pan with nonstick cooking spray.
3. Add butter, brown sugar, cinnamon, ¼ teaspoon nutmeg and cloves to bowl with bread mixture; beat with electric mixer at low speed about 1 minute or until smooth. Add apple, dried fruit and egg; beat until blended. Stir in nuts. Pour into prepared pan.
4. Bake 1 hour 15 minutes to 1 hour 30 minutes or until toothpick inserted into center comes out clean. Cool in pan 10 minutes; remove to wire rack to cool slightly. Serve warm; top with whipped cream and additional nutmeg, if desired.

TRIPLE LEMON CAKE

MAKES 9 TO 12 SERVINGS

CAKE

- 1¾ cups all-purpose flour
- 1¼ teaspoons baking powder
- ½ teaspoon salt
- 4 eggs, separated
- 1½ cups granulated sugar
- ¾ cup (1½ sticks) butter, softened
- Grated peel of 2 lemons (1 tablespoon)
- ¼ cup lemon juice

LEMON CURD

- 1 cup granulated sugar
- ¾ cup (1½ sticks) butter
- ⅔ cup lemon juice
- Grated peel of 2 lemons (1 tablespoon)
- ⅛ teaspoon salt
- 5 eggs, beaten
- ¼ cup whipping cream

TOPPING

- 1 package (8 ounces) cream cheese, softened
- 1½ cups powdered sugar
- Lemon peel strips (optional)

1. Preheat oven to 350°F. Spray 9-inch square baking pan with nonstick cooking spray.
2. For cake, combine flour, baking powder and salt in medium bowl; mix well. Beat 4 egg whites in large bowl with electric mixer at high speed until stiff peaks form. Transfer to small bowl.
3. Beat 1½ cups granulated sugar and ¾ cup butter in large bowl with electric mixer at medium speed until light and fluffy. Add 4 egg yolks, one at a time, beating well after each addition. Add 1 tablespoon lemon peel and ¼ cup lemon juice; beat until well blended. Beat in flour mixture at low speed just until blended. Gently stir in half of egg whites. Fold in remaining egg whites until no streaks of white remain. Spread batter in prepared pan.
4. Bake 35 to 38 minutes or until toothpick inserted into center comes out clean. Cool in pan on wire rack 10 minutes.
5. Meanwhile, for lemon curd, combine 1 cup granulated sugar, ¾ cup butter, ⅔ cup lemon juice, 1 tablespoon lemon peel and ⅛ teaspoon salt in medium saucepan; cook over medium heat until butter is melted and sugar is dissolved, stirring frequently. Gradually whisk in beaten eggs in thin, steady stream.

Cook over medium-low heat 5 minutes or until thickened to consistency of pudding, whisking constantly. Strain through fine-mesh sieve into medium bowl. Remove ½ cup lemon curd to small bowl. Press plastic wrap onto surface of remaining lemon curd; cool to room temperature. Refrigerate until cold and thickened.

6. Stir cream into reserved ½ cup lemon curd. Poke holes all over warm cake with skewer. Pour cream mixture over cake, spreading to cover surface and pressing mixture into holes. Cover and refrigerate 1 hour.

7. For topping, beat cream cheese in large bowl with electric mixer at medium speed 1 minute or until creamy. Add powdered sugar and 1 cup lemon curd; beat 2 minutes or until well blended and fluffy. Spread remaining lemon curd over top of cake. Gently spread topping over lemon curd layer. Refrigerate 2 hours or overnight. Garnish with lemon peel strips.

GINGER PEAR COBBLER

MAKES 8 TO 10 SERVINGS

- 7 firm ripe d'Anjou pears (about 3½ pounds), peeled and cut into ½-inch pieces
- ⅓ cup packed brown sugar
- 1 cup plus 2 tablespoons all-purpose flour, divided
- 2 tablespoons lemon juice
- 2 teaspoons ground ginger, divided
- ½ teaspoon ground cinnamon
- ⅛ teaspoon ground nutmeg
- ¼ cup granulated sugar
- 1½ teaspoons baking powder
- ¼ teaspoon salt
- ¼ cup (½ stick) cold butter, cut into small pieces
- ¼ cup whipping cream
- 1 egg, lightly beaten
- Sparkling or coarse sugar (optional)

1. Preheat oven to 375°F. Spray 9-inch square baking dish with nonstick cooking spray.
2. Combine pears, brown sugar, 2 tablespoons flour, lemon juice, 1 teaspoon ginger, cinnamon and nutmeg in large bowl; toss to coat. Spoon into prepared baking dish.
3. Combine remaining 1 cup flour, 1 teaspoon ginger, granulated sugar, baking powder and salt in medium bowl; mix well. Add butter; mix with fingertips until shaggy clumps form. Add cream and egg; stir just until combined. Drop topping, 2 tablespoonfuls at a time, into mounds over pear mixture. Sprinkle with sparkling sugar, if desired.
4. Bake 40 to 45 minutes or until filling is bubbly and topping is golden brown.

CHOCOLATE GUINNESS CAKE

MAKES 10 TO 12 SERVINGS

- 1 package (about 15 ounces) chocolate cake mix
- 1 package (about 3 ounces) instant chocolate pudding and pie filling mix
- 4 eggs
- 1 cup Guinness stout
- ¾ cup vegetable oil
- ½ cup semisweet chocolate chips
- Chocolate Guinness Glaze (recipe follows)
- Whipped cream or vanilla ice cream (optional)
- Grated semisweet chocolate (optional)

1. Preheat oven to 350°F. Grease and flour 12-inch (10-cup) bundt or tube pan.
2. Combine cake mix and pudding mix in large bowl; mix well. Whisk eggs, Guinness and oil in medium bowl until blended. Add to cake mix; beat 2 minutes or until well blended. Stir in chocolate chips. Pour batter into prepared pan.
3. Bake 55 minutes or until toothpick inserted near center comes out clean. Meanwhile, prepare Chocolate Guinness Glaze.
4. Cool cake in pan 5 minutes; invert onto serving plate. Brush half of glaze over cake while still warm. Let cake cool completely before drizzling with remaining glaze. Serve cake with whipped cream, if desired; garnish with grated chocolate.

CHOCOLATE GUINNESS GLAZE

Combine 1 cup (2 sticks) butter, 1 cup sugar, ½ cup Guinness and ¼ cup water in large saucepan; cook and stir over medium heat until butter is melted and sugar is dissolved, stirring occasionally. Remove from heat; stir in 3 ounces chopped bittersweet chocolate until melted and smooth. Cool to room temperature before using.

STICKY TOFFEE PUDDING

MAKES 6 SERVINGS

CAKE

- 1 cup pitted chopped dates (5 ounces)
- ⅔ cup water
- ½ teaspoon baking soda
- 1 cup all-purpose flour
- 1 teaspoon baking powder
- ¼ teaspoon salt
- ¾ cup packed brown sugar
- 5 tablespoons butter, softened, plus additional for greasing ramekins
- 1 egg
- ½ teaspoon vanilla

TOFFEE SAUCE

- ¼ cup (½ stick) butter
- ½ cup packed brown sugar
- ½ cup whipping cream
- ½ teaspoon vanilla
- ⅛ teaspoon salt

1. Preheat oven to 350°F. Butter six 4-ounce ramekins; place on parchment paper-lined baking sheet.
2. Combine dates and water in medium saucepan; bring to a boil over medium-high heat. Remove from heat; stir in baking soda. Set aside to cool.
3. Combine flour, baking powder and ¼ teaspoon salt in small bowl; mix well. Beat ¾ cup brown sugar and 5 tablespoons butter in large bowl with electric mixer at medium-high speed about 3 minutes or until light and fluffy. Scrape down side of bowl. Add egg and ½ teaspoon vanilla; beat about 1 minute or until well blended. Stir in date mixture with spatula until blended. Add flour mixture; stir just until blended. Divide batter evenly among prepared ramekins.
4. Bake 25 to 30 minutes or until toothpick inserted into centers comes out clean. Meanwhile, prepare sauce. Melt ¼ cup butter in medium saucepan over medium heat. Add ½ cup brown sugar, cream, ½ teaspoon vanilla and ⅛ teaspoon salt; cook over medium-high heat about 5 minutes or until brown sugar is dissolved and sauce is reduced to 1 cup, stirring frequently. Remove from heat; cover to keep warm.

5. Poke tops of puddings with skewer at ½-inch intervals. (Make sure to poke holes all the way through to the bottom.) Gradually pour half of sauce over puddings; let stand about 15 minutes or until all of sauce is absorbed. Run sharp knife around edge of ramekins; invert onto serving plates. Reheat remaining sauce over medium-low heat; pour over tops of puddings. Serve immediately.

PORTER CAKE

MAKES 10 SERVINGS

- 3½ cups all-purpose flour
- 1½ teaspoons pumpkin pie spice
- 1 teaspoon baking powder
- ½ teaspoon salt
- 1 cup (2 sticks) butter
- 1 bottle (10 ounces) porter or stout
- 1 cup packed brown sugar
- 1½ cups golden raisins
- 1½ cups raisins
- Finely grated peel of 1 orange
- 2 eggs, lightly beaten
- ¼ chopped candied citrus peel
- ¼ cup candied cherries

1. Preheat oven to 350°F. Grease 9-inch springform pan; line bottom with parchment paper. Grease parchment paper; dust bottom and side of pan with flour, tapping out excess. Line baking sheet with foil.
2. Combine 3½ cups flour, pumpkin pie spice, baking powder and salt in large bowl; mix well. Combine butter, porter and brown sugar in large saucepan; cook over medium heat about 7 minutes or until butter is melted and brown sugar is dissolved, stirring occasionally. Remove from heat; stir in raisins and orange peel. Let cool about 15 minutes or until just warm.
3. Add porter mixture and eggs to flour mixture; stir just until combined. Fold in candied citrus peel and cherries. Pour batter into prepared pan; place on prepared baking sheet.
4. Bake 60 to 65 minutes or until toothpick inserted into center comes out clean. Cool in pan on wire rack 15 minutes; remove side of pan and cool completely on wire rack.

Berry Shortcake Trifles

MAKES 4 SERVINGS

Lemon Curd

- 1 cup granulated sugar
- ⅔ cup lemon juice
- ½ cup (1 stick) butter
- 1 tablespoon grated lemon peel
- ¼ teaspoon salt
- 4 eggs, beaten

Berries

- ½ pound fresh strawberries, stemmed and chopped
- ½ pound fresh blueberries
- ⅓ cup granulated sugar

Whipped Cream

- 4 ounces cream cheese, softened
- 6 tablespoons powdered sugar, divided
- 1 cup whipping cream, divided
- ¼ teaspoon vanilla
- 1 prepared pound cake (about 14 ounces), cut into ½-inch cubes

1. Combine 1 cup granulated sugar, lemon juice, butter, lemon peel and salt in medium saucepan; cook and stir over medium heat until butter is melted and sugar is dissolved. Gradually whisk in eggs in thin, steady stream. Cook over medium-low heat 5 minutes or until thickened to consistency of pudding, whisking constantly. Strain through fine-mesh sieve into medium bowl. Press plastic wrap onto surface; refrigerate at least 2 hours or until cold.
2. Meanwhile, combine strawberries, blueberries and ⅓ cup granulated sugar in medium bowl; stir gently to blend. Cover and refrigerate at least 2 hours.
3. Beat cream cheese, 3 tablespoons powdered sugar, 2 tablespoons cream and vanilla in large bowl with electric mixer at medium speed 3 minutes or until smooth. Beat remaining cream and 3 tablespoons powdered sugar in medium bowl with electric mixer at high speed until stiff peaks form. Fold whipped cream into cream cheese mixture until blended.
4. Drain berries, reserving liquid. For each serving, place ½ cup pound cake cubes in bottom of wide-mouth 1-pint jar; sprinkle with 1 tablespoon reserved berry liquid. Top with scant ¼ cup whipped cream mixture, 2 tablespoons lemon curd and 1 tablespoon berries. Repeat layers. Refrigerate overnight.

TOFFEE CAKE WITH WHISKEY SAUCE

MAKES 9 SERVINGS

- 8 ounces chopped dates
- 2¼ teaspoons baking soda, divided
- 1½ cups boiling water
- 2 cups all-purpose flour
- ½ teaspoon salt
- ¾ cup (1½ sticks) butter, softened
- ½ cup granulated sugar
- ½ cup packed dark brown sugar
- 2 eggs
- 1 teaspoon vanilla
- 1½ cups butterscotch sauce
- 2 tablespoons whiskey
- 1 cup glazed pecans* or chopped toasted pecans
- Vanilla ice cream

***Glazed or candied pecans may be found in the produce section of the supermarket with other salad toppings, or they may be found in the snack aisle.**

1. Preheat oven to 350°F. Spray 9-inch square baking pan with nonstick cooking spray.
2. Combine dates and 1½ teaspoons baking soda in medium bowl. Stir in boiling water; let stand 10 minutes to soften. Mash with fork or process in food processor until mixture forms paste.
3. Combine flour, remaining ¾ teaspoon baking soda and salt in medium bowl; mix well. Beat butter, granulated sugar and brown sugar in large bowl with electric mixer at medium speed 3 minutes or until creamy. Add eggs, one at a time, beating well after each addition. Beat in vanilla. Add flour mixture alternately with date mixture; beat at low speed just until blended. Spread batter in prepared pan.
4. Bake about 30 minutes or until toothpick inserted into center comes out with moist crumbs. Cool in pan on wire rack 15 minutes.
5. Pour butterscotch sauce into medium microwavable bowl; microwave on HIGH 30 seconds or until warm. Stir in whiskey. Drizzle sauce over each serving; sprinkle with pecans and top with ice cream.

Glazed Lemon Loaf

MAKES 8 TO 10 SERVINGS

Cake

- 1½ cups all-purpose flour
- ½ teaspoon baking powder
- ½ teaspoon baking soda
- ½ teaspoon salt
- 1 cup granulated sugar
- 3 eggs
- ½ cup vegetable oil
- ⅓ cup lemon juice
- 2 tablespoons butter, melted
- 1 teaspoon lemon extract
- ½ teaspoon vanilla

Glaze

- 3 tablespoons butter
- 1½ cups powdered sugar
- 2 tablespoons lemon juice
- 1 to 2 teaspoons grated lemon peel

1. Preheat oven to 350°F. Grease and flour 8×4-inch loaf pan.
2. For cake, combine flour, baking powder, baking soda and salt in large bowl; mix well. Whisk granulated sugar, eggs, oil, ⅓ cup lemon juice, 2 tablespoons melted butter, lemon extract and vanilla in medium bowl until well blended. Add to flour mixture; stir just until blended. Pour batter into prepared pan.
3. Bake about 40 minutes or until toothpick inserted into center comes out clean. Cool in pan 10 minutes; remove to wire rack to cool 10 minutes.
4. Meanwhile, prepare glaze. Melt 3 tablespoons butter in small saucepan over medium-low heat. Whisk in powdered sugar, 2 tablespoons lemon juice and 1 teaspoon lemon peel; cook until smooth and hot, whisking constantly. Pour glaze over warm cake; smooth top. Cool completely before slicing. Garnish with additional 1 teaspoon lemon peel, if desired.

Irish Stout Caramel Milkshake

MAKES 2 SERVINGS

Stout Caramel Sauce (recipe follows)

1 pint (2 cups) vanilla ice cream, softened slightly

½ cup whole milk

1. Prepare Stout Caramel Sauce; refrigerate until ready to use.
2. Combine ice cream, milk and ¼ cup caramel sauce in blender; blend 30 to 40 seconds or until smooth and frothy.
3. Pour into two serving glasses; drizzle with additional caramel sauce.

Stout Caramel Sauce

MAKES ABOUT 1⅓ CUPS SAUCE (ENOUGH FOR 4 BATCHES OF MILKSHAKES)

1 bottle (about 11 ounces) Irish stout

1 cup sugar

1 cup whipping cream, warmed until hot

1. Pour stout into medium saucepan; cook over medium heat until reduced to about ½ cup.
2. Stir in sugar; continue cooking until mixture registers 275°F on candy thermometer. Remove from heat; carefully whisk in cream about ⅓ cup at a time. Let cool to room temperature.
3. Transfer to airtight container; cover and refrigerate at least 1 hour to thicken and cool completely.

IRISH COFFEE

MAKES 1 SERVING

- 6 ounces freshly brewed strong black coffee
- 2 teaspoons packed brown sugar
- 2 ounces Irish whiskey
- ¼ cup whipping cream

Combine coffee and brown sugar in Irish coffee glass or mug. Stir in whiskey. Pour cream over back of spoon into coffee.

LAST WORD

MAKES 1 SERVING

- ¾ ounce gin
- ¾ ounce green Chartreuse
- ¾ ounce maraschino liqueur
- ¾ ounce lime juice
- Lime twist

Fill cocktail shaker with ice; add gin, Chartreuse, liqueur and lime juice. Shake until blended; strain into chilled coupe or cocktail glass. Garnish with lime twist.

Frozen Mudslide

MAKES 1 SERVING

1 cup vanilla ice cream
1 ounce vodka
1 ounce coffee liqueur
1 ounce Irish cream liqueur
1 to 2 tablespoons whipping cream or half-and-half (optional)
Chocolate syrup (optional)
Whipped cream and mini chocolate chips (optional)

1. Combine ice cream, vodka, coffee liqueur and Irish cream liqueur in blender; blend until smooth. Add cream, if desired, to reach desired consistency.
2. If desired, garnish glass with chocolate syrup swirls before pouring drink into glass. Hold glass at 90-degree angle; gently squeeze chocolate syrup onto side of glass while turning glass. Or squeeze syrup in vertical lines up and down side of glass.
3. Pour drink into prepared glass; garnish with whipped cream and chocolate chips.

VOLUME MEASUREMENTS (dry)

1/8 teaspoon = 0.5 mL
1/4 teaspoon = 1 mL
1/2 teaspoon = 2 mL
3/4 teaspoon = 4 mL
1 teaspoon = 5 mL
1 tablespoon = 15 mL
2 tablespoons = 30 mL
1/4 cup = 60 mL
1/3 cup = 75 mL
1/2 cup = 125 mL
2/3 cup = 150 mL
3/4 cup = 175 mL
1 cup = 250 mL
2 cups = 1 pint = 500 mL
3 cups = 750 mL
4 cups = 1 quart = 1 L

VOLUME MEASUREMENTS (fluid)

1 fluid ounce (2 tablespoons) = 30 mL
4 fluid ounces (1/2 cup) = 125 mL
8 fluid ounces (1 cup) = 250 mL
12 fluid ounces (1 1/2 cups) = 375 mL
16 fluid ounces (2 cups) = 500 mL

WEIGHTS (mass)

1/2 ounce = 15 g
1 ounce = 30 g
3 ounces = 90 g
4 ounces = 120 g
8 ounces = 225 g
10 ounces = 285 g
12 ounces = 360 g
16 ounces = 1 pound = 450 g

DIMENSIONS

1/16 inch = 2 mm
1/8 inch = 3 mm
1/4 inch = 6 mm
1/2 inch = 1.5 cm
3/4 inch = 2 cm
1 inch = 2.5 cm

OVEN TEMPERATURES

250°F = 120°C
275°F = 140°C
300°F = 150°C
325°F = 160°C
350°F = 180°C
375°F = 190°C
400°F = 200°C
425°F = 220°C
450°F = 230°C

BAKING PAN SIZES

Utensil	Size in Inches/Quarts	Metric Volume	Size in Centimeters
Baking or Cake Pan (square or rectangular)	8×8×2	2 L	20×20×5
	9×9×2	2.5 L	23×23×5
	12×8×2	3 L	30×20×5
	13×9×2	3.5 L	33×23×5
Loaf Pan	8×4×3	1.5 L	20×10×7
	9×5×3	2 L	23×13×7
Round Layer Cake Pan	8×1½	1.2 L	20×4
	9×1½	1.5 L	23×4
Pie Plate	8×1¼	750 mL	20×3
	9×1¼	1 L	23×3
Baking Dish or Casserole	1 quart	1 L	—
	1½ quart	1.5 L	—
	2 quart	2 L	—